4
Topics in Organometallic Chemistry

Organometallic Bonding and Reactivity

Fundamental Studies

Volume Editors: J.M. Brown and P. Hofmann

With contributions by

P.B. Armentrout, D. Braga, A. Dedieu, P. Gisdakis, A. Görling, F. Grepioni,
F. Maseras, N. Rösch, S.B. Trickey

Springer

Chemistry Library

The series *Topics in Organometallic Chemistry* presents critical overviews of research results in organometallic chemistry, where new developments are having a significant influence on such diverse areas as organic synthesis, pharmaceutical research, biology, polymer research and materials science. Thus the scope of coverage includes a broad range of topics of pure and applied organometallic chemistry. Coverage is designed for a broad academic and industrial scientific readership starting at the graduate level, who want to be informed about new developments of progress and trends in this increasingly interdisciplinary field. Where appropriate, theoretical and mechanistic aspects are included in order to help the reader understand the underlying principles involved.
The individual volumes are thematic and the contributions are invited by the volumes editors.
In references Topics in Organometallic Chemistry is abbreviated *Top. Organomet. Chem.* and is cited as a journal.

Springer WWW home page: http://www.springer.de

ISSN 1436-6002
ISBN 3-540-64253-6
Springer-Verlag Berlin Heidelberg New York

Library of Congress Cataloging–in–Publication Data
Organometallic bonding and reactivity : fundamental studies / volume editors, J. M. Brown and P. Hofmann ; with contributions by P. B. Armentrout ... [et al.].
    p.   cm. – – (Topics in organometallic chemistry ; 4)
    Includes bibliographical references.
    ISBN 3-540-64253-6 (hardcover : alk. paper)
    1. Organometallic compounds. 2. Chemical bonds. 3. Reactivity (Chemistry)   I. Brown, John M.   II. Hofmann, P. (Peter), 1947– III. Series.
QD411.5.072   1999
547'.05––dc21
                                                                99-33293
                                                                CIP

© Springer-Verlag Berlin Heidelberg 1999
Printed in Germany

The use of general descriptive names, registered names, trademarks, etc. in this publication does not imply, even in the absence of a specific statement, that such names are exempt from the relevant protective laws and regulations and therefore free for general use.

Cover: Friedhelm Steinen-Broo, Pau/Spain; MEDIO, Berlin
Typesetting: Data conversion by MEDIO, Berlin

SPIN: 10543767   66/3020 - 5 4 3 2 1 0 – Printed on acid-free paper.

Volume Editors

Dr. John M. Brown
Dyson Perrins Laboratory
South Parks Road
Oxford OX1 3QY,
E-mail: john.brown@chemistry.oxford.ac.uk

Prof. Peter Hofmann
Organisch-Chemisches Institut
Universität Heidelberg
Im Neuenheimer Feld 270
D-69120 Heidelberg, Germany
E-mail: ph@phindigo.oci.uni-heidelberg.de

Editorial Board

Preface

General

The making and breaking of carbon-metal bonds is fundamental to all the processes of organometallic chemistry and moreover plays a significant role in homogeneous as well as heterogeneous catalysis. This rather blunt statement emphasises the extent to which a proper understanding of the structure, energetics and reactivity of C–M bonds is at the core of the discipline. In order to accept it, a proper definition of the terms involved is required. Quite simply we define the metal-carbon bond in its broadest sense to embrace carbon linked to transition-metals, lanthanides and actinides, and main group metals. We do not distinguish between formally covalent single or multiple bonding on the one hand and η-bonding on the other. In the studies to be described in the following chapters, the emphasis will be on transition metal complexes and insofar as the fundamentals come under scrutiny, simple metal alkyls or related species (metal alkenyl, alkynyl, aryl, or allyl) will play an emphatic part. The central role of metal alkyls and their congeners and especially the role of their metal carbon linkage in homogeneous catalysis may be appreciated by considering some key reaction steps leading to their formation or breakdown. There follows a few prominent examples of transition metal mediated stoichiometric or catalytic processes:

- In homogeneous hydrogenation of double bonds, the stepwise reaction of an η^2-coordinated alkene with dihydrogen gives first an alkyl metal hydride, and then the decoordinated alkane by elimination.
- In the heterogeneous catalysis of hydrogenation, surface-bound metal alkyls play a pivotal role in the reaction cycle.
- Homogeneous or heterogeneous dehydrogenation reactions of hydrocarbons involve transition metal alkyl hydrides, which may undergo ß-elimination and decoordination of H_2 and an alkene.
- In hydroformylation, a metal alkyl is formed in similar manner but intercepted by *cis*-ligand migration to coordinated CO; the reductive elimination then involves an acyl metal hydride.
- Hydrosilylation and many more related addition reactions of X–H or X–Y units to unsaturated organic substrates proceed via metal alkyl (or alkenyl, aryl) intermediates, which are produced by insertion steps into M–H or M–X,Y bonds. Hydrocyanation of alkenes and dienes figures prominently in this context.

- For transition metal catalysed alkene amination, a process of great industrial potential, the most promising catalytic cycles are based upon the intermediacy of alkyl metal complexes, formed either by amine addition to a metal-coordinated olefin or by olefin insertion into M–H and M–N bonds, respectively. Metal catalysed alkyne amination and hydration reactions are related cases.
- For the simplest mechanism of alkene polymerisation the alkyl chain grows through an alkyl migration to coordinated alkene; the same process is responsible for C–C bond formation in alkene dimerisations and oligomerisations.
- The copolymerisation of alkenes and CO to 1,4-polyketone polymers involves successively a palladium alkyl and acyl, the sequence being continued by migration of the acyl to η^2-coordinated alkene, and further cis-ligand migration to coordinated CO.
- In metathesis and ROMP polymerisation, the key steps are a template cycloaddition between metal alkylidene and alkene, leading to metal alkyl bonds in a metallacyclobutane structure, and the reverse process with opposite regioselectivity.
- Palladium and nickel-catalysed cross-couplings involve successive addition of a carbon electrophile and a carbon nucleophile to the metal and then an elimination of cis-adjacent alkyl groups; for the related Heck reaction the key step is the cis-ligand migration of a palladium alkyl of electrophilic origin to a coordinated alkene.
- The catalytic amination or carboxyalkylation of halogenated arenes as well as the catalytic arylation of carbonyl compounds using palladium catalysts create aryl metal intermediates en route to the C–N bond forming elimination step.
- Intermediates of olefin oxidation reactions of the Wacker-type are hydroxy-substituted metal alkyls of e.g. palladium.
- Metal η^3-allyls, often in equilibrium with their η^1-allyl isomers, have a broad base of catalytic involvement best appreciated through the exometallic reaction of cationic palladium allyls with nucleophiles or the intermediacy of allyl nickel complexes in hydrocyanation of butadiene. The chemistry of η^3-benzyl systems is related.
- Migration of an unsaturated alkyl group from iron to carbon is the basis of the most convincing explanation for Fischer-Tropsch telomerisation.
- C–H activation of alkanes, a fundamental step for C–H functionalization reactions in both chemical and biological systems gives a metal alkyl as the first formed intermediate. C–H functionalization reactions of alkenes and arenes, e.g. hydrovinylation or the Murai reaction and related processes, involve metal aryls or alkenyls en route to functionalized hydrocarbons.
- Last but not least, numerous stoichiometric reactions of reactants where the transition metal acts as a template, permit the chemo- and stereoselective synthesis of complex organic molecules through intermediates with M–C bonds.

Naturally this constitutes an incomplete list. Given the enormously broad scope of known or potential transformations of organic substrates involving M–C bonds, how then may the experimentalist or theoretician contribute to basic understanding? Here it is convenient to separate the contributions of these two communities, although in practice there is considerable convergence of effort.

Experimental Studies

On the structural side there is an accumulating body of results from X-ray, neutron and electron diffraction invaluable for developing a systematic corpus of data on bond lengths and bond angles, and defining the trends with respect to variation of metal and co-ligands. Fast, highly efficient X-ray instrumentation in the form of area detector, CCD, rotating anode and synchrotron technology has brought about a revolution in speed for the determination of molecular structures of even the largest organometallic systems in the solid state. Dunitz, Bürgi and others made seminal contributions to our knowledge of solid state structure/reactivity relationships. Now a large body of X-ray and neutron diffraction data is quickly and easily retrievable from structural databases, and can be widely used to "map out" parts of energy surfaces or of specific reaction pathways or to derive subtle variations of molecular structure from large series of related compounds. The accuracy of X-ray data permits answers to questions about the nature of C–M bonding versus Van der Waals contacts. Taken together, information from diffraction experiments form the basis of efforts to tailor the structure of organometallic compounds ("ligand design") for specific functions in organometallic chemistry and catalysis. Solid state structure determination provides the theme for the Chapter by Braga and Grepioni "Static and Dynamic Structures of Organometallic Molecules and Crystals".

Despite the high level of precision of contemporary solid state structural studies, more detailed information on energetics and reactivity patterns need to be collected from other experimental sources. Two areas of current endeavour provide significant results.

Mass spectrometric techniques, which are elaborated in Armentrout's Chapter "Gas Phase Organometallic Chemistry", possess the power to provide direct information on the energetics of transient species generated in the gas-phase. Recent reports have shown, that gas phase investigations of reaction pathways and energetics are feasible even for "real" catalytically active complexes, as for C–H activating $[Cp^*Ir(PR_3)]$ 14-electron intermediates, for Grubbs type $(PR_3)_2Cl_2Ru(carbene)$ olefin metathesis and $Cp_2Zr(R)^+$ olefin polymerisation catalysts. Armentrout's Chapter is largely concerned with guided ion beam tandem MS, and other workers have applied FT Ion Cyclotron Resonance [FTICR]. By analysis of the kinetic energy release distribution, experimental bond energies may be derived, and compared with the predictions of increasingly sophisticated calculations. Much of the mass spectrometric work involves bare metal cations (or metal oxide cations MO^+) and permits direct comparisons of chemoselectivity, regioselectivity and reactivity. For example, the reaction of light metal cations with hydrocarbons can result in some C–C cleavage in competition with C–H activation. Heavier transition metal cations lead to dehydrogenation via C–H activation. MS experiments may be extended to ligated metal ions; a significant reaction between $ScMe_2^+$ and cycloalkanes is a sigma-bond metathesis occurring by a four-centre transition state, in competition with dehydrogenation so that a range of R_2Sc^+ species is observed. Interesting proposals of a "two-state-reactivity" have been employed to explain the gas phase reactivity of MO^+ fragments with organic substrates. Important questions concerning the transferability of gas phase reaction patterns to solution chemistry remain to be answered.

In catalysis it is a familiar truism that many of the most interesting species are highly elusive; their short lifetime under normal reaction conditions precludes detection. Time-resolved IR spectroscopy holds considerable promise for the definition of species in this category. Given a spectrometer with 200 femtosecond time resolution, intermediates of very short lifetime may be detected provided that their transient concentration is sufficient. Laser photolysis of the $Tp^*Rh(CO)_2$ complex at 295 nm occurs with a high quantum yield (0.3) for CO dissociation and C–H activation from hydrocarbon solvent. This lies in contrast to the quantum yield of 0.01 for the corresponding Cp^* complex. In the period of 500 ns after CO dissociation, several intermediates are observed. First a molecular alkane complex ensues, which dissociates one of the pyrazole units over 200 ps. The dissociated species undergoes first C–H insertion and then rechelation of the pyrazole, both on a 200 ns timescale, to give the stable C–H activation product. The energy barrier for the critical C–H insertion is around 35 kJmol^{-1}. The fast IR approach is made more powerful when coupled to classical mechanistic probes. In a related instance where $Cp^*Ir(PMe_3)$ is the coordinatively unsaturated fragment, the existence of an alkane complex en route to the C–H activation/insertion product was proved by the photolysis of alkylhydride isotopomers and satisfactory correlation of the results with a kinetic model requiring an alkane complex.

Photoelectron spectroscopy is another important experimental tool which has provided deeper insight into bonding patterns and electronic structures of organometallic compounds and into M–C interactions. Here – in contrast to simple organic molecules – one observes the breakdown of Koopmans' theorem. This inevitably necessitates either the spectroscopic comparison of series of related and specifically modified model compounds, or the use of appropriate computational procedures in order to identify the nature of observed ionisation events. These can then be related to a qualitative or quantitative bonding description of the species in question. A large body of PE spectroscopic information on organometallics has been collected in the past, but surprisingly its direct influence and use as a guideline for synthesis and its impact for expanding mechanistic knowledge and devising novel structures or reaction pathways has been somewhat limited. Certainly further effort will be very important here.

Modern spectroscopic techniques also provide intimate details of the structure of surface bound groups. For example, high-resolution electron energy loss spectroscopy (HREELS) provides an equivalent IR spectrum of adsorbent which can be compared with theoretical calculation (DF calculations).

Solution thermochemistry should be mentioned as a further area of fundamental studies, which are of great importance, because they can provide reference data for estimating reaction enthalpies or for establishing useful additive and incremental schemes for energy calculations of single step organometallic reactions or catalytic cycles. Unfortunately, solid and reliable thermochemistry data for organometallic reactions in condensed phase are rather scarce, and only a few groups are operating seriously in this field. Their results form an important link to the results of theoretical calculations and may serve as a credibility nexus between theory and experiment.

Theoretical Studies

All types of fundamental experimental studies of organometallic structures, structural dynamics, energetics and reactivity in the solid state, in solution or in the gas phase are intimately connected to theoretical chemistry with its large body of modern computational tools. It is certainly adequate to state, that during the last 10 to 15 years we have witnessed a dramatic change of the role that is played by theoretical chemistry for organometallic chemistry and catalysis research. The rapid development of computers and of programming technology and the concomitant commercial availability or free accessibility of often easy-to-handle, graphics- and screen-oriented program packages have caused a revolutionary change in attitudes towards theory among organometallic chemists. The 1998 Nobel Prize in Chemistry was awarded to two of the pioneers of theoretical and computational chemistry, John A. Pople and Walter Kohn and nicely testifies to this statement. The experimental chemist has access to most levels of theory, ranging from molecular mechanics approaches and semiempirical quantum chemistry to highly sophisticated, correlated density functional and *ab initio* (molecular orbital, valence bond) calculations.

For this reason most organometallic and catalysis research laboratories have come to use quantum chemical calculations on a routine basis during the past 10 years. It is interesting – and to some extent surprising – to realise that the employment of theoretical methods either for analysing experimental results or to plan organometallic molecular structure and function is an even more routinely established tool in industrial R&D labs engaged in organometallic or catalysis research, than in academic laboratories. Contemporary quantum chemistry allows one to perform calculations not only for small model systems, from which basic electronic structure patterns and unifying concepts can be derived, but also allows modelling of real systems. Models of bonding and electronic structure, based upon more qualitative or semi-quantitative concepts and methods like ligand and crystal field theory, the angular overlap model, PMO theory and orbital interaction rules, all variants of Extended Hückel-type calculations and their descriptive one-electron MO theory tools for molecular or extended systems are useful tools for analysing and understanding many features of electron-

ic structure, bonding and reactivity. Computational chemistry with *first principle ab initio* or density functional methods make a reliable numerical assessment of structures and (relative) energies increasingly feasible, however. Em-
bedding methods, combining *ab initio* or density functional quantum chemistry
for selected substructures with an appropriate force field or semiempirical MO
treatment of the ligand environment extend the utility of the basic methods. The
range of theoretical techniques available is complete when quantum dynamics
studies and the computational modelling of solvent effects are included.

Density functional methods, developed on the basis of the Hohenberg-Kohn
and the Kohn-Sham theorems have been very successfully for molecular quantum chemistry during the last decade. The main attraction lies in their ability to
treat even rather large molecules with comparable accuracy but more easily,
faster and thus more cost-effectively than by standard wave function based
methods. DF routines are implemented in, and can be conveniently used within,
most of the standard *ab initio* program packages. The chapter by Görling, Trickey, Gisdakis and Rösch "A Critical Assessment of Density Functional Theory
with Regard to Applications in Organometallic Chemistry" gives a descriptive,
detailed and critical survey of the theoretical background, the history and the
power of DF methods, drawing attention also to their inherent limitations. The
essence of the more widely used DF approximations is described and the authors
emphasise caveats as well as offering perspectives of the Kohn-Sham (KS) theory
for molecular quantum chemistry. KS orbitals and KS eigenvalues are discussed
and their relationship to the Hartree-Fock (HF) description of electronic structure is presented in a nicely transparent and elucidating manner. The concept of
functionals and the various types of local, approximate gradient-corrected and
hybrid functionals used in DF calculations are explained in an appropriate way
for a chemistry oriented, non-specialist readership. A balanced view of the treatment of exchange and correlation phenomena by DF methods is presented and
is followed by a section, which provides a concise and highly informative body
of data and references allowing a quantitative calibration and validation of DF
results in comparison to those from conventional *first principles* wave function
based quantum chemical methods. A critical evaluation of the general performance of DF calculational methods for organometallic systems and the presentation of case studies of organometallic oxo systems and their reactions (OsO_4 olefin dihydroxylation, CH_3ReO_3 olefin epoxidation) complete this chapter. From
the viewpoint of the experimentalist who is interested in understanding or applying DF calculations for his own research, this is complementary to and more
chemically oriented (less mathematical) than most other fundamental reviews
or books.

The theme of organometallic reactivity as treated by quantum chemical calculations is continued in Dedieu's chapter "Theoretical Treatment of Organometallic Reaction Mechanisms and Catalysis" where at the beginning a general
overview is given of the theoretical "toolbox" of methods currently in use for
treating organometallic reactions, ranging from qualitative molecular orbital
theory to *ab initio*, density functional, combined quantum chemical/molecular

mechanics (QM/MM) and molecular dynamics simulation (QM/MD, e.g. Carr-Parinello) with their merits and shortcomings. The main body of this Chapter deals with selected examples of homogeneous catalytic processes which are of great industrial interest. The author first addresses in depth the topic of early transition metal (Ti, Zr) metallocene based olefin polymerisation, in particular with respect to the mechanistic significance and the requirements for a correct theoretical description of agostic M–C–H interactions. A rather detailed review of quantum dynamics simulation studies is given. Like this Chapter as a whole, it is intended to provide the reader with not just numerical computational results, but also qualitative interpretations and general concepts derived from theoretical findings. Dedieu's second case study is linked to the first, as it also deals with olefin polymerisation catalysis. Here representative quantum chemical studies of a more recent generation of catalyst systems, based upon late transition metal (Ni, Pd) diimine complexes, are outlined and discussed. The importance and influence of solvent effects, not taken into account by most quantum chemical studies of organometallic structure and reactivity, is considered in the last section of Dedieu's chapter. Possible theoretical approaches to solvent effects are collected from the literature and from the author's own research, olefin hydroformylation and the Wacker process being chosen as examples. An extensive reference list of theoretical work on organometallic reactions and catalytic cycles completes the Chapter.

There is a strong current impetus from the introduction of hybrid quantum mechanics/molecular mechanics methods, which permit calculations on large and realistic molecular systems and reaction pathways without resorting to truncated models, where hydrogen atoms replace actual organic substituents of e.g. large ligands (e.g. PH_3 stands for P^tBu_3 etc.). Such structural simplifications remain meaningful and acceptable only if general features of electronic structures and qualitative, transferable classifications of organometallic structure and reactivity are required. If, however, steric interactions or the precise tailoring of stereoelectronic effects play a decisive role in chemo- and stereoselectivity, particularly in the field of enantioselective transition metal catalysis, realistic models have to be used in computational studies, and the QM/MM methodology offers the chance to do so. In the Chapter by Maseras "Hybrid Quantum Mechanics/Molecular Mechanics Methods in Transition Metal Chemistry" the reader can get an instructive first-hand introduction into this rapidly expanding field of computational chemistry. The author has actively participated in the development of the Integrated Molecular Orbital Molecular Mechanics (IMOMM) method, which is one of the presently available QM/MM approaches, as they now are already implemented in many quantum chemistry program packages. An appreciable part of this Chapter is devoted to an introduction into the methodological features of QM/MM models, incorporating enough qualitative description and explanation of the theoretical background to make the approach easily comprehensible without going too deeply into mathematical formalisms. The main part of Maseras' contribution focusses upon applications, and starts with three structural studies of sterically highly congested transition metal complexes, allowing the reader to develop a feeling for the reliability of QM/MM (IMOMM) results which are shown to be useful for separating steric and electronic effects upon structure and reactivity. The specific advantages of hybrid QM/MM techniques become clearly visible in the theoretical description of transition metal catalysed olefin polymerisation, the mechanistic and energetic details of which can be compared to the analysis given in Dedieu's chapter. Other examples chosen by Maseras are the asymmetric dihydroxylation of olefins by osmium tetroxide, where the detailed analysis of the author's QM/MM study can be compared to the result from other computational strategies as outlined in the Chapter by Görling, Trickey, Gisdakis and Rösch. The importance of steric prerequisites for agostic interactions in organometallic systems, and hence the need for a complete incorporation of steric effects in theoretical treatment of compounds where agostic interactions play an important role, is emphasised and made clear. Generally, it is shown, how the role of steric bulk upon the stability of organometallic molecular geometries can be adequately dealt with in a qualitative and even a quantitative way by use of QM/MM methods. The perspective for treating large bioinorganic complexes is outlined in computational model studies of porphyrin complexes, for which comparison and evaluation of different theoretical approaches is given.

This volume "Organometallic Bonding and Reactivity: Fundamental Studies" of the series "Topics in Organometallic Chemistry" presents a survey by renowned experts of important experimental and theoretical developments to understand basic aspects of bonding, energetics, reactivity, molecular geometries and solid-state structures of organometallic compounds. We are grateful to the authors for their cooperation, for sharing their expertise and for communicating results of their own and of others, which provide a fascinating overview of the situation at the frontiers of the disciplines treated in this volume.

Oxford, August 1999   John M. Brown
Heidelberg, August 1999   Peter Hofmann

Contents

Gas-Phase Organometallic Chemistry

Peter B. Armentrout

e-mail: armentrout@chemistry.utah.edu

Department of Chemistry, University of Utah, Salt Lake City, UT 84112, USA

Studies of organometallic chemistry in the gas phase can provide substantial quantitative information regarding the interactions of transition metals with carbon centers. In this review, the techniques associated with such studies are outlined with an emphasis on guided ion beam tandem mass spectrometry. The use of this technique to measure thermodynamic information is highlighted. Periodic trends in covalent bonds between first, second and a few third row transition metals and small carbon ligands are discussed and shown to correlate with a carefully defined promotion energy. The bond energies for dative interactions between the first row transition metal ions and ethene, benzene and alkanes are also reviewed. With this thermochemical background, the reactions of atomic transition metal ions with alkanes (methane, ethane and propane) are reviewed and periodic variations in the reactivity are highlighted. An overview of our results on the effects of ancillary ligands (CO and H_2O) and oxo ligands on the reactivity of transition metal centers are then provided.

Keywords: Mass spectrometry, Transition metal ions, Thermochemistry, Bond activation, Bond energies

Topics in Organometallic Chemistry, Vol. 4
Volume Editors: J.M Brown and P. Hofmann
© Springer-Verlag Berlin Heidelberg 1999

List of Abbreviations

AE	appearance energy
BDE	bond dissociation energy
CM	center-of-mass
E_e	energy of an electron
EI	electron ionization
ICR	ion cyclotron resonance
IE	ionization energy
KERD	kinetic energy release distribution
REMPI	resonance enhanced multiphoton ionization
rf	radio frequency
RRKM	Rice-Ramsperger-Kassel-Marcus
TS	transition state
SI	surface ionization

1
Introduction

Why study organometallic chemistry in the gas phase? One of the principle driving forces behind such studies is to provide a fundamental view of catalytic reactions that occur at transition metal sites. Yet it is nearly impossible to carry out practical catalysis in the gas phase [1–4] because the concentrations involved are pitifully low and it is difficult to obtain the number of reactant encounters needed to observe meaningful turnover ratios without competing side reactions and loss mechanisms. Precisely because of this, however, the gas-phase environment provides a place where a detailed study of the intrinsic reactivity of a metal site with a reagent of interest, unfettered by the influences of solvent and unrestricted by the demands of stability imposed by the 18-electron rule can be carried out. Such details, which are difficult to obtain in condensed phase systems, include (1) quantitative thermodynamic data of metal–substrate bonding; (2) characterization of electronic state and spin effects on reactivity; (3) comprehensive periodic trends (vertical and horizontal) in reactivity; (4) specific mechanistic insight into reaction pathways; and (5) systematic studies of the effects of oxidation state, selective ligation, and solvation on reaction energetics and mechanisms. These features of gas-phase chemistry are highlighted in the present review. In addition, insights are provided into studies of how reactivity changes with the number of metal centers, i.e. a gas-phase approach to heterogeneous catalysis. Such metal cluster studies are an active component of research in our laboratories [5–7] but are beyond the scope of this review.

In addition to providing insight into the chemistry that is observed in condensed phase systems, there is also the prospect of observing different reactivities than those typical of such media. For instance, one of the early results of gas-phase chemistry was the observation that atomic metal ions could activate the unstrained C–C bonds of alkanes [8]. Comparable processes are rarely observed in condensed phase media despite intense efforts. If we can understand what allows such unique transformations to be facile in the gas phase, the prospects of engineering a true catalyst should be enhanced.

Another way of viewing the role that gas-phase studies can play in understanding organometallic chemistry is the realization that the active reagents in homogeneous catalysis are coordinatively unsaturated transition metal–ligand complexes. While ligand field theory allows us to organize a tremendous amount of information regarding stable organometallic compounds, the open shell character of the unsaturated complexes makes them less easily organized by these closed shell 18-electron rules. Although such stable complexes are the starting materials in organometallic reactions and homogeneous catalysis, the key reactive intermediates are these unsaturated transition metal–ligand complexes that have an open site of reactivity on the metal center formed by loss of one or more ligands from the stable reagent. Precisely because they are reactive (and therefore good catalysts), such intermediates are transient and difficult to study. Little is known about the thermodynamics of such reactive species, in contrast to organic chemistry, where it is straightforward to look up typical C–C, C–H, C–O, etc. bond energies. This enables fairly accurate predictions of the overall thermochemistry and thus the feasibility of virtually any reaction desired. Although it is impossible to characterize the bond energies of all possible ligands with all possible unsaturated metal complexes, gas-phase thermochemistry does provide a quantitative handle on the strength of these interactions and the effects of adjoining ligands on these interactions.

2
Experimental Methods

The gas-phase chemistry of transition metal ions has been examined by many experimental techniques. Two complementary methods have dominated the research: ion cyclotron resonance (ICR) mass spectrometry and ion beam mass spectrometry. Other techniques that have been used include high-pressure mass spectrometry, flowing afterglow methods [9] and collisional activation by tandem mass spectrometry [10]. This review will focus on the abilities of and results from ion beam mass spectrometry, the method used in my laboratory.

2.1
Atomic Metal and Metal Complex Ion Sources

Atomic metal ions can be produced using a variety of methods: electron ionization (EI) of volatile organometallic complexes, laser vaporization (or ablation) of bulk metal samples, a glow discharge where the metal of interest is the cathode, surface ionization (SI, also called thermionic emission) of metal salts and organometallic complexes, and resonance-enhanced multiphoton ionization (REMPI) of organometallic complexes and gas-phase metal atoms. These various methods differ greatly in the distribution of electronic states produced. As any ionization process is an energetic one, it is likely that excited electronic states of transition metals (which have numerous low-lying levels) are produced. As a consequence, the experimental generation of specific states of transition metal ions is difficult and a distribution of states is generally produced. The low-lying states of transition metals involve only s and d orbitals and therefore are metastable. Optical transitions between these states are parity forbidden such that radiative relaxation does not occur on the time scale of most experiments [11].

Organometallic complexes of transition metal cations can also be formed using a variety of techniques. Electron ionization and multiphoton ionization of stable organometallic precursors can be employed or, alternatively, chemical reactions of atomic metal ions with suitable reagents can be used to form many complexes. A key question in such methods is the internal (vibrational, rotational, and electronic) energy of the complexes. As a consequence, the coupling of atomic metal ion sources with a high-pressure reaction region for complex generation is a particularly advantageous approach and one that we have utilized extensively in our laboratories. This is discussed further below.

2.1.1
Electron Ionization

In an electron ionization (EI) source, high energy electrons impinge on a volatile organometallic compound leading to dissociation and ionization. The efficiency of this process depends on the energy of the electron (Ee), which clearly must exceed some threshold for production of the atomic metal ion, called the appearance energy (AE) of the atomic metal ion. To provide sufficient intensity, the Ee must generally be substantially above the AE. A number of experiments have documented that excited state populations can exceed 50% for most transition metal ions formed under such conditions [12–24]. The extent of excitation does not vary substantially once the Ee is 10–20 eV above the AE.

It is interesting that early experiments failed to find evidence for excited states of transition metal ions formed by EI, but the reasons for this are subtle. Researchers wrongfully assumed that low-lying states would be short-lived, while they are actually metastable (see above). They expected to see differences in reactive rates or product branching ratios for metal ions in different electronic states; however, for exothermic reactions (the most easily observed), the differences in reactivities between states are often small.

As noted above, EI sources can provide a straightforward means of creating organometallic complexes given a suitable precursor. For example, we have generated $Fe(CO)_5^+$ ions using such a source [25]. However, even when the electron energy is set to a value very close to the ionization energy of $Fe(CO)_5$, the molecular ions produced have considerable internal energy. Hence, the use of such sources without the addition of a means to thermalize the ions is inappropriate for quantitative studies of their chemistry and thermodynamics.

2.1.2
Laser Vaporization and Glow Discharge

Laser vaporization and glow discharge sources are intense means for generating atomic metal ions but produce a multitude of excited states [26, 27]. Strong prima facie evidence for this is the observation of multiply charged atomic ions [26, 29–31], species that are even more energetic than excited states of the corresponding singly charged ions. However, coupling of such sources with high-pressure devices (such as a flow tube, see below) make these sources particularly powerful experimental tools. At present, there is no quantitative work characterizing the detailed state populations produced under various laser irradiation or discharge conditions. This is largely because such studies are difficult and the results would depend on a host of experimental conditions.

2.1.3
Surface Ionization

One of the best controlled sources for producing atomic metal ions is the surface ionization (SI) source. In such a source, a rhenium or tungsten filament is heated to 1800–2300 K and exposed to the vapor of an organometallic compound or a metal salt. Decomposition occurs and atoms with low ionization energies desorb from the filament surface with a probability described by the Saha-Langmuir equation [32–34]. Consistent with this, there is presently good evidence that the populations of electronic states produced are characteristic of the filament temperature [23, 35]. Because the available energy is low ($k_B T$=0.2 eV at 2300 K), only the lowest energy states are formed. This is not a particularly intense source (10^5–10^7 ions s^{-1}), but it is extremely stable.

2.1.4
Multiphoton Ionization

The most precise means of creating a well-defined electronic state of transition metal ions is the resonance-enhanced multiphoton ionization (REMPI) source, as implemented by Weisshaar [36, 37]. Unfortunately, this source is also the most difficult (and most expensive) to implement experimentally. Comparison of the state-specific results obtained with this source are generally in good agreement with other techniques [38] and provide an absolutely vital check of the state characterization of these other methods.

2.1.5
High-Pressure Sources

The EI, laser vaporization, and glow discharge sources are intense sources of metal ions but yield many excited states. One way to eliminate the excited states produced in these sources is to quench them by introducing a high-pressure gas into the source region. We have demonstrated that nearly pure beams of the ground states of many transition metal ions can be produced in this way by coupling a flow tube source with EI [39], laser vaporization [27], and glow discharge sources [25]. Studies indicate that thousands of collisions with a species like Ar or CH_4 can be needed to efficiently remove the excited states [19]. A typical ICR experiment, where the ions are produced directly in the cell, has difficulty attaining this number of collisions, but flow tubes have a sufficiently high pressure (0.5–1 Torr) of He or Ar that 10^4–10^5 collisions are always present. Further, additional reagents (e.g. we have used O_2, NO, and CH_4) can be added to the flow to enhance the quenching.

Such high-pressure sources are also capable of producing organometallic complexes by reaction of the atomic metal ions with suitable reagents or by three-body condensation. This technique has permitted us to generate large numbers of unsaturated organometallic complexes without the need for a stable

organometallic precursor [40]. Further, the large number of thermalizing collisions cool the internal energies of the complexes. Although there are no unam-
biguous probes of the temperatures of such complexes, all evidence indicates
that the internal energies are well characterized as having reached equilibrium
with the bath gas temperature [41–43].

2.2
Mass Spectrometric Methods

2.2.1
ICR Mass Spectrometry

Ion cyclotron resonance (ICR) and its derivative technique, Fourier transform
ICR (FT-ICR or FTMS) [44], uses crossed electric and magnetic fields to trap
ions for further study. This permits the time dependence of the ion population
to be measured, typically over a range of 10^{-3} to 10^{1} s. Thus, rates, $k(T)$, for ion-
molecule reactions at ambient conditions (generally room temperature) and
product branching ratios of these reactions are routinely measured using this
technique. One powerful aspect of ICR methods is that sequential reactions can
be monitored easily and reaction pathways identified with double-resonance
techniques.

Both EI and laser vaporization sources have been used routinely in ICR experiments. While SI sources have been used for production of alkali metal ions
[45], the ion intensities of transition metal ions are too small for this to be a
workable ICR source for these species (although attempts have been made). It is
technically possible to perform REMPI in an ICR cell but this has not yet been
utilized for transition metal ion studies. The more extensive use of external ion
sources for ICRs now allows all of the sources discussed above to be used with
ICRs (including glow discharges).

2.2.2
Ion Beam Mass Spectrometry

Ion beam mass spectrometers, the instruments used in our laboratory [46], involve an ion source, a mass spectrometer used to select the ionic reactant, a reaction zone, a second mass spectrometer to analyze ionic products, and an ion
detector. The reaction zone is designed so that reactions occur over a well-defined path length and at a pressure low enough that all products are the result of
single ion-neutral encounters. The sensitivity of the detectors is high; in our
case, sufficient that individual ions can be detected with near 100% efficiency.
An important variant of such instruments is a "guided" ion beam mass spectrometer, which utilizes an octopole ion beam guide, first developed by Teloy
and Gerlich [47, 48], in the reaction region. This device utilizes eight rods ar-
ranged in an octagonally symmetric array around the ion beam path. Alternate
phases of a radio frequency (rf) electric potential are applied to alternate rods to

create a potential well in the direction perpendicular to the axis of the rods. This potential well traps product ions and unreacted metal ions keeping them confined until they drift from the reaction region where they are accelerated, mass analyzed, and detected. This greatly enhances the sensitivity of the experiment and permits routine measurement of the distribution and absolute zero of the ion kinetic energy.

In an ion beam experiment, one must convert from the laboratory to the center-of-mass (CM) energy scale. While the kinetic energy of the reactant ion is measured in the laboratory frame, some of this energy is tied up in the kinetic energy of the center-of-mass of the reactant system through the laboratory. Because the total mass of the system does not change during reaction, conservation of linear momentum demands that this fraction of the total energy remains constant. Therefore, it is not available to induce chemical reactions. The energy that is available for chemistry is called the center-of-mass energy, $E(CM)$, and is easily calculated (in the stationary target limit) as $E(CM)=E(lab) m/(M+m)$ where m and M are the masses of the neutral and ionic reactants, respectively. The stationary target assumption does not include the motion of the neutral reactant molecules or the distribution in the ion beam energy. Rather, the effects of these distributions are explicitly included in analysis of the data, as described in Sect. 3.1.2.

In addition to converting from laboratory to CM energies, the measured intensities of the product and reactant ions must be converted into an absolute reaction cross section, $\sigma(E)$ [46]. This conversion requires knowledge of the neutral reactant pressure and the length of the interaction region, both quantities that are straightforward to measure. The cross section can be thought of as the effective area that the reactants present to one another such that they collide and then proceed to the desired products. Typical units are 10^{-16} cm^2=Å^2, properly reflecting the size of the reactants. Cross sections are a direct measure of the probability of the reaction at a given kinetic energy and are directly related to a microcanonical rate constant by $k(E)=v\,\sigma(E)$ where v is the relative velocity of the reactants, $v=(2E/\mu)^{1/2}$, and μ is the reduced mass of the reactants. A temperature-dependent rate constant, $k(T)$, can be obtained by integrating $k(E)$ over a Maxwell-Boltzmann distribution of relative velocities (although the distribution of internal energies of the reactants is generally not in equilibrium with their kinetic energies).

The primary reason for the development of ion beam technology is that the kinetic energy of the reactant ion can be varied easily over a wide range simply by changing the voltage difference between where the ions are formed and where they react. This ability provides the key difference between ion beam and ICR technology, although acceleration of ions using resonant excitation is being increasingly pursued in ICR mass spectrometers [49–51]. The energy range accessible in ion beam instruments extends from energies as low as thermal, $3k_B T/2$ (300 K)=0.03 eV=3 kJ mol^{-1}, to hundreds of volts, 100 eV=$3k_B T/2$ at 750,000 K. Although the organometallic chemistry most relevant to condensed phase systems occurs in the thermal energy regime, hyperthermal kinetic energies can induce the making and breaking of chemical bonds that cannot occur under thermal conditions. This enables us to study the activation of many bonds, to probe the potential energy surfaces and thus the mechanisms of reactions, and to examine endothermic reactions, thereby acquiring thermodynamic information.

Because the ion source is physically separated from the interaction region, ion beam mass spectrometry can use any of the sources described above. To date, electron ionization, laser vaporization, glow discharge, surface ionization, and high-pressure variations of these sources have been coupled with ion beam instruments. This versatility enables excellent control of the electronic and internal energies of the transition metal reactant cations.

3
Thermochemistry of Metal–Carbon Bonds

One of the key abilities of guided ion beam mass spectrometry is the ability to examine reactions at hyperthermal conditions, thus allowing endothermic reactions to be studied. By analyzing the kinetic energy dependence of particular reactions, the energy thresholds for reaction can be measured and related back to specific bond energies of interest. We have applied this methodology to a wide number of systems in order to measure periodic trends in transition metal–carbon bond energies. Such information has long been available for first row transition metals and has been reviewed several times previously, most recently by Armentrout and Kickel [52] who systematically reevaluated all of our previous work. These data along with thermodynamic information from other laboratories have been tabulated by Freiser [53]. Despite some overlap with previous reviews [52, 54–56], it is worth introducing the primary concepts here in order to utilize this information in understanding the mechanisms and potential energy surfaces discussed in later sections. In addition, we compile here for the first time reasonably complete information for second row transition metal cations. Although some of these values are still preliminary, trends in the thermochemistry are quite clear and help guide our discussion of the differences in reactivity down the periodic table.

In this and later sections, energy will be given in either kilojoules per mole, kJ mol^{-1}, or electron volts, eV. The latter are the natural units for our experiments. The conversion between these units and the passé kcal mol$^-$ is 1 eV= 96.49 kJ mol^{-1}=23.06 kcal mol^{-1}.

3.1
Methods of Analysis

3.1.1
Reactions

In many of our studies, the schematic reaction (1) is used to determine the thermochemistry of M^+–L bond energies.

$$M^+ + RL \rightarrow ML^+ + R \tag{1}$$

A simple example is formation of metal-methyl cations from ethane (L=R= CH_3). A slightly more complex case is the generation of methyl-methylidene cations (L=CH_2) which can be formed from methane (R=H_2) or cyclopropane (R= C_2H_4). In all these cases, the threshold for reaction, E_0, is related to the desired bond energy by Eq. (2).

$$D_0(M^+-L) = D_0(R-L) - E_0(1) \tag{2}$$

Thermochemistry for neutral species can be determined in a related fashion by examination of reaction (3).

$$M^+ + RL \rightarrow ML + R^+ \tag{3}$$

Here the species R is chosen to have a low ionization energy (IE), such that reaction (3) competes effectively with reaction (1). An example of interest here is formation of metal-methyls from *neo*-pentane [L=CH_3, R=$C(CH_3)_3$]. The appropriate thermochemistry is obtained using Eq. (4).

$$D_0(M-L) = D_0(R-L) + IE(R) - IE(M) - E_0(3) \tag{4}$$

Another general type of reaction that we have used to derive thermodynamic data is collision-induced dissociation (CID), reaction (5) where Rg is an inert collision gas. We usually use Xe for reasons described elsewhere [57–59]. This reaction provides thermodynamic information straightforwardly as the threshold for reaction equals the desired bond energy, $E_0 = D_0(M^+-L)$.

$$ML^+ + Rg \rightarrow M^+ + L + Rg \tag{5}$$

However, accurate thermodynamic information is obtained only when the analysis of the CID cross sections includes the effects of multiple ion-neutral collisions and the lifetime for dissociation. The first effect is handled by extrapolating our data to zero neutral pressure, rigorously single collision conditions [25, 60]. The second effect becomes more important as the metal-ligand species become larger and more complex, which can eventually lead to a lifetime for dissociation that is comparable to the time it takes the ions to travel from the collision cell to the detector ($\sim 10^{-4}$ s). This leads to a delay in the observed onset for dissociation, a "kinetic shift". We account for this effect by using Rice-Ramsperger-Kassel-Marcus (RRKM) theory [61] to calculate a dissociation probability as a function of the ion internal energy [62, 63].

In all these systems, accurate thermochemistry for the products is obtained only if the reactions have no barriers in excess of the endothermicities of the reactions studied. In contrast to the situation ordinarily found in condensed phases, the assumption of no reverse activation barriers is often a reasonable one for ion-molecule reactions because of the strong long-range ion-induced dipole potential [64]. The most obvious illustration of this fact is that exothermic ion-molecule reactions are generally observed to proceed without an activation energy. The converse must also be true, endothermic ion-molecule reactions generally proceed once the available energy exceeds the thermodynamic threshold [65]. We have explicitly tested this assumption in several reactions where the thermochemistry is well established [43, 66–71] although the observation of the true thermodynamic threshold can require extremely good sensitivity [66]. Exceptions do occur, however, and can result from spin or orbital angular momentum conservation restrictions [65, 72], or the presence of a tight transition state (TS) along the reaction coordinate, as illustrated below for some C–H bond activation steps [73–76]. For CID reactions of organometallic complexes, quantum mechanics demonstrate that there should be no reverse activation energies, a consequence of the heterolytic bond cleavages involved [77], although dissociation to an excited state asymptote can occur [25].

The accuracy of the thermochemistry measured in such experiments can be experimentally verified by using more than one single reaction system. Although this is not possible in all cases, such checks have been performed for many of the systems described here. Alternatively, the thermochemistry can be verified by comparison with values from other experiments [78–85] and ab initio theory [86–94].

3.1.2
Data Analysis

Thresholds (E_0 values) for reactions (1), (3) and (5) are determined by detailed modeling of the experimental cross sections using a mathematical expression justified by theory [95, 96] and experiment [65, 68–71, 97–99]. This model is given by Eq. (6).

$$\sigma(E) = \sigma_0 \sum g_i (E + E_i - E_0)^n / E \tag{6}$$

Here, σ_0 is a scaling factor, E is the relative kinetic energy, and n is an adjustable parameter. The sum is over the contributions of individual reactant states (vibrational, rotational and/or electronic), denoted by i, with energies E_i and populations g_i ($\Sigma g_i = 1$) The use of this equation to analyze the threshold behavior of reaction cross sections makes the statistical assumption that the internal energy of the reactants is available to effect reaction. It also presumes that the energy dependence (as determined by n and σ_0) does not vary with the state i (although this latter assumption can be inaccurate for different electronic states).

Before comparison with the experimental data, the model of Eq. (6) is convoluted over the explicit distributions of the kinetic energy of the neutral and ion reactants, as described previously [46, 100, 101]. The σ_0, n, and E_0 parameters are then optimized by using a non-linear least-squares analysis to give the best reproduction of the data. Because Eq. (6) includes all sources of energy for the reactants, the thresholds and bond energies obtained using these methods correspond to 0 K thermochemistry. Uncertainties in E_0 reflect the range of threshold values obtained for different data sets with different values of n and the error in the absolute energy scale. In cases where the internal energy of the reactants is appreciable and the vibrational frequencies are not well established, the uncertainty also includes variations in the calculated internal energy distribution of the reactants, E_i. Such uncertainties can also influence the lifetimes for dissociation in CID experiments. The accuracy of the thermochemistry obtained by this modeling procedure is dependent on a variety of experimental parameters that have undergone extensive discussion [52, 54, 65].

3.2
Covalent Metal–Carbon Bonds

3.2.1
Cations

Bond energies for first and second row transition metal cations with CH_3, CH_2 and CH are given in Table 1. Figure 1 illustrates the periodic trends in these val-

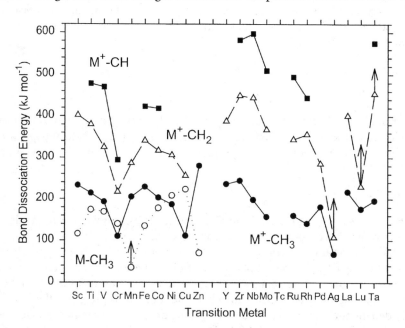

Fig. 1. Periodic trends in the bond energies (in kJ mol^{-1}) of first, second and selected third row transition metal cations with CH_3 (*solid circles*), CH_2 (*open triangles*) and CH (*closed squares*) and of neutral first row transition metals with CH_3 (*open circles*)

Table 1. Transition Metal–Carbon Bond Energies (0 K) in kJ mol^{-1}.[a]

M	M^+–CH_3	M^+–CH_2	M^+–CH	M–CH_3	M^+–$(CH_3)_2$
Sc	233(10)	402(23)		116(29)*	464(5)
Ti	214(3)	380(9)	478(5)	174(29)*	472(25)
V	193(7)	325(6)	470(5)	169(18)*	391(7)
Cr	110(4)	217(4)	294(29)	140(7)	
Mn	205(4)	286(9)		>35(12)	
Fe	229(5)	341(4)	423(29)[b]	135(29)*	409(12)
Co	203(4)	317(5)	420(37)[c]	178(8)	
Ni	187(6)	306(4)		208(8)	
Cu	111(7)	256(5)		223(5)	
Zn	280(7)			70(10)	
Y	236(5)	388(13)			
Zr	244(15)[d]*	449(5)[d]*	582(13)[d]*		
Nb	198(28)[e]*	444(3)[e]*	597(23)[e]*		
Mo	157(12)[e]*	329(12)[e]*	509(10)[e]*		
Ru	160(6)[f]	344(5)[f]	494(15)[f]		
Rh	141(6)[g]	356(8)[g]	444(12)[g]		
Pd	181(10)[h]	285(5)[h]			
Ag	67(5)[i]	>107(4)[i]		134(7)[i]	
La	217(15)	401(7)			
Lu	176(20)	>230(6)			
Ta	196(3)[j]*	>454[j]*	575(9)[j]*		

[a]Values are taken from [52] unless otherwise noted. [b]Hettich RL, Freiser BS (1986) J Am Chem Soc 108:2537. [c][75]. [d][145]. [e][146]. [f][148]. [g][154]. [h][155]. [i][156]. [j][152].
*Preliminary values not yet thoroughly evaluated

ues. Clearly, for a specific metal, M^+–CH bonds are stronger than M^+–CH_2 bonds, which are stronger than M^+–CH_3 bonds. This simply reflects the triple, double and single bond character of these bonds. Across the periodic table, double-humped trends, common to many transition metal properties, are observed. These are easily rationalized. In the first row, the metal ions having the weakest bond energies are Cr^+ and Cu^+, which have $3d^5$ and $3d^{10}$ electron configurations. Formation of a bond with any transition metal requires decoupling the bonding electrons on the metal from the remaining electrons. Because of the stability of the half-filled and filled shells of Cr^+ and Cu^+, the energetic costs of this decoupling are larger than for other metal ions and hence the bond energies are smaller. In the second row, Ag^+ ($4d^{10}$) behaves the same way, while Mo^+ ($4d^5$) is less strongly inhibited.

We have previously noted [102, 103] that the periodic trends in these bond energies can be quantified by correlating them with the promotion energy, Ep, defined as the energy required to prepare a metal ion in a particular electronic configuration suitable for bonding. For convenience, the promotion energies we use here are those calculated by Carter and Goddard [104]. This correlation is shown in Fig. 2. In the case of the M–C single bonds, Ep is for promoting to a $s^1 d^n$ configuration where the s electron is spin-decoupled from the electrons in the d orbitals. Clearly, almost all of the M^+–CH_3 bond energies correlate nicely with this promotion energy and the correlation is very similar for first, second and the few available third row metal cations. This indicates that there is considerable s character in the metal-bonding orbital. On the basis of these results, the "intrinsic" metal–carbon single bond energy, i.e. the bond energy expected for a metal prepared to bond strongly to a methyl group, is about 240 kJ mol^{-1} (the y-axis intercept).

The two exceptions to this good correlation are $PdCH_3^+$ and $AgCH_3^+$, the two metals having the highest promotion energies, 344 and 518 kJ mol^{-1}, respectively.

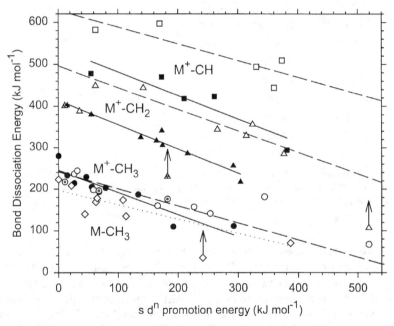

Fig. 2. Transition metal cation ligand bond energies (in kJ mol^{-1}) vs the atomic metal ion promotion energy to a $s^1 d^n$ spin-decoupled state (see text). Results are shown for CH_3 (*circles*), CH_2 (*triangles*) and CH (*squares*) with data taken from Table 1. Diamonds show results for neutral metal–CH_3 bond energies. *Closed symbols* show results for first row metal cations, *open symbols* for second row metal cations, and *dotted symbols* for third row metal cations. Linear regression fits to the data (excluding $PdCH_3^+$, $AgCH_3^+$ and YCH_2^+, see text) are shown by *solid lines* for first row metal complexes, by *dashed lines* for second row metal complexes, and by the *dotted line* for neutral M–CH_3 complexes.

In the case of Pd$^+$ ($4d^9$), the high cost of promoting to the $5s^14d^8$ state means that it is more favorable to form a bond between the methyl group and the $4d$ σ-orbital, even though this bond is not as strong as that with the $5s$ orbital. Calculations [91] confirm this result and also indicate that other late second row transition metal cations (Ru and Rh, which also have among the higher Ep values) should also have considerable d character in their bonding orbital. In the case of Ag$^+$ ($4d^{10}$), calculations [91] indicate that no promotion takes place and the bond is simply a one-electron bond involving the methyl group and the empty $5s$ orbital.

Similar correlations are found between promotion energies and the M$^+$–CH$_2$ and M$^+$–CH bond energies. In these cases, the promotion energies shown are to a s^1d^n state where the s and one (for a double bond) or two (for a triple bond) d electrons are decoupled from the remaining electrons. The correlations are all nearly parallel with one another, but there is a striking difference in the magnitudes of the correlations for first and second row metals, i.e. in the intrinsic bond energies. It is clear from Fig. 2 that the M$^+$–CH$_2$ and M$^+$–CH bond energies are stronger for the second row transition metals than for the first row metals. Apparently, the average π-bond energy for first row metals is about 150 kJ mol^{-1} while it is about 200 kJ mol^{-1} for the second row metals. The YCH$_2{}^+$ bond energy (Ep=36 kJ mol^{-1}) deviates from this pattern because it contains more $4d^2$ than $5s^14d^1$ character [105]. Theory also finds that RuCH$_2{}^+$ and RhCH$_2{}^+$ utilize primarily $4d^{n+1}$ rather than $5s^14d^n$ metal character in the bonding; however, the correlation with s^1d^n promotion energies is still quite good, perhaps indicating that the sigma bond involves more s character than the calculations suggest. The enhanced bonding for the second row metals can be attributed to a better spatial overlap of the $4d$ and $5d$ orbitals with those of carbon than the $3d$ orbitals and to smaller exchange energies. The electronic ground state configurations of all first and second row metals are the same except for the cases of Fe and Co vs Ru and Rh, where the first row metal methylidene cations have high-spin ground states (quartet and triplet, respectively) while the second row species have low-spin ground states (doublet and singlet, respectively) [105].

3.2.2
Bis-Ligated Cations

Key intermediates in the reactions of transition metal cations with alkanes are alkyl hydrides, R–M$^+$–H, and bis-alkyls, R–M$^+$–R', as discussed in Sect. 4. Therefore, a quantitative understanding of these reactions requires knowledge of both the first and second covalent bonds to the transition metal. Unfortunately, the second bond energies are not as easily measured experimentally, although this has been accomplished for several dimethyl species, as listed in Table 1. In addition, D_0[Sc$^+$–(H)(CH$_3$)]=479±5 [106] and D_0[V$^+$–(H)(CH$_3$)]=390±16 kJ mol^{-1} [107] have been measured. In all cases listed, there are good reasons to believe that the information listed pertains to species containing two covalent metal ligand bonds. In other cases, CoC$_2$H$_6{}^+$ and NiC$_2$H$_6{}^+$, thermodynamic information obtained and attributed to dimethyl species [108] is more likely to pertain to an ethane adduct structure [52, 109]. For $M(CH_3)_2^+$ species where M=Sc, Ti, V or Fe (Table 1), the sum of the two covalent bond dissociation energies (BDEs) in these species is 2.0±0.2 times as large as the single covalent M^+–CH_3 BDEs. We have previously found that the sum of the $M(CH_3)_2^+$ BDEs correlates reasonably well with the Ep for formation of two covalent bonds [102], i.e. the same Ep as for M^+–CH_2 BDEs. The intrinsic dimethyl BDE of this correlation is 480±20 kJ/mol, twice the intrinsic value for a single metal ion-methyl BDE, 240 kJ mol^{-1}.

More information on the bis-ligated transition metal cations is available from theoretical studies. Rosi et al. [110] have examined the transition metal dimethyl cations of the first and second row series and concluded that the ethane adduct structure is favored by Cr^+, Mo^+, Cu^+ and Ag^+ (the ions having half-filled and filled d shells), while Sc^+, Y^+, Ti^+, Zr^+ and Nb^+ (all early transition metal ions) should have the dimethyl structures. For all other metal ions, the two structures are close in energy. Blomberg et al. [111] examined the methyl hydrides of the second row transition metal cations (and neutrals). They found that only $HYCH_3^+$ and $HZrCH_3^+$ were more stable than the M^++CH_4 species and that when M=Ru, Rh or Pd, there is no minimum associated with the $HMCH_3^+$ structure. Similar calculations on the first row $HMCH_3^+$ species by Hendrickx et al. [112] found parallel results: $HScCH_3^+$ is the only species more stable than M^++CH_4, and Fe, Co, Ni and Cu have no minima for the $HMCH_3^+$ geometry. Similar conclusions have been reached for $HCoC_2H_5^+$ [113, 114] and $HCoC_3H_7^+$ [114].

3.2.3
Neutrals

Similar trends and correlations with Ep are also observed for neutral transition metal–carbon bonds although the data base is more limited. Figure 1 shows periodic trends in the bond energies of the first row transition metal methyl neutrals. These experimental values are in good agreement with theoretical results [90], except for the case of $ScCH_3$, which is a tentative value that has not been rigorously analyzed. MCH_3 bond energies exhibit the same kind of periodic trends as the M^+–CH_3 bond energies but displaced by one position in the periodic table. However, the trends are not comparable for isovalent neutral and cationic species, i.e. those having the same number of valence electrons (e.g. V and Cr^+ or Ni and Cu^+). Rather the neutral metals having the lowest bond energies are Mn and Zn, which have stable $4s^23d^5$ and $4s^23d^{10}$ electron configurations. This emphasizes the fact that bonding is controlled by the detailed electronic structure of the metal center, i.e. Cr^+ ($3d^5$) is more comparable to Mn ($4s^23d^5$) than it is to V ($4s^23d^3$) and Cu^+ ($3d^{10}$) is like Zn ($4s^23d^{10}$) rather than Ni ($4s^13d^9$).

The periodic trends in the neutral metal–carbon bond energies can also be correlated by a promotion energy to a s^1d^n configuration where the s electron is spin-decoupled from the largely nonbonding d electrons [54, 102, 103]. The correlation is similar to that for the MCH_3^+ bond energies (Fig. 2). The intrinsic bond energy obtained is 213 kJ mol^{-1}, somewhat less than that obtained for the cations. Comparison with cationic and neutral metal hydrides indicates that this difference can be attributed to a small electrostatic contribution to the M^+–CH_3 bond energies [54, 102, 103].

3.3
Dative Metal–Carbon Bonds

We have also examined the binding of several stable molecules to transition metal cations. These include the carbon-containing species CO [25, 62, 115–119], C_2H_4 [120], C_6H_6 [121] and alkanes [109, 122–126], as well as other ligands such as water [127] and ammonia [128]. Our work on metal–carbonyl bond energies has been reviewed thoroughly [40] and so this discussion will not be repeated here. Instead, we concentrate on a brief examination of the bonding of transition metal cations to ethene, benzene and the alkanes.

3.3.1
Ethene

We recently measured the binding energies of the first row transition metal cations (Ti^+–Cu^+) with one and two ethene molecules [120]. The results are listed in Table 2 and shown in Fig. 3. In several cases, $Fe^+(C_2H_4)$, $V^+(C_2H_4)_2$, and $Mn^+(C_2H_4)_2$, the values have been corrected for dissociation to an excited state asymptote. These corrections are rationalized in detail in our original paper. The results are in reasonable agreement with several theoretical calculations and a host of experiments [120].

Table 2. Transition metal–ligand bond energies (0 K) in kJ mol^{-1}

L	C_2H_4[a]		C_6H_6[b]		CH_4	
M	M^+–L	LM^+–L	M^+–L	LM^+–L	M^+–L	LM^+–L
Ti	146(11)		259(9)	253(18)		
V	124(8)	127(14)	234(10)	246(18)		
Cr	96(11)	108(11)	170(10)	232(18)		
Mn	91(12)	88(15)	133(9)	203(16)		
Fe	145(11)	152(16)	207(10)	187(16)	57(3)[c]	97(4)[c]
Co	186(9)	152(14)	256(11)	167(14)	90(6)[d]	96(5)[d]
Ni	182(11)	173(15)	243(11)	147(12)		
Cu	176(14)	174(13)	218(10)	155(12)		

[a][120]. [b][121]. [c][109]. [d][124].

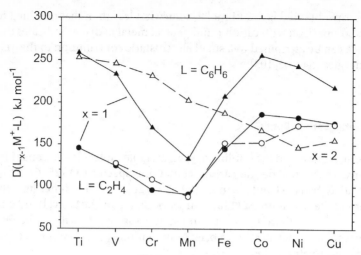

Fig. 3. First (*closed symbols*) and second (*open symbols*) bond energies (in kJ mol^{-1}) of ethene (*circles*) and benzene (*triangles*) to the first row transition metal cations

Clearly the first and second bond energies are similar to one another and show the same periodic trends. In addition, these values are very similar to metal cation–CO bond energies with the exception of Mn$^+$(CO) and Mn$^+$(C$_2$H$_4$). This is reasonable as both CO and C$_2$H$_4$ ligands are σ-donor–π-acceptor types of ligands. The weakest bond energies are for Mn$^+$, the only metal ion where the 4*s* orbital is occupied in the complex [129]. Occupation of this orbital leads to relatively severe repulsion with the two-electron donating ligand.

3.3.2
Benzene

Results for the bonding of one and two benzene molecules to the first row transition metal ions (Ti$^+$–Cu$^+$) are also listed in Table 2 and shown in Fig. 3 [121]. The values for the first benzene ligand are in reasonable agreement with theory [130] and other experimental values. Periodic trends in these values parallel those for the ethene ligand. The early metal ions (Ti, V, Cr) have benzene bond energies that are about 1.8 times those of the ethene bond energies, while the factor is only 1.4 for the late metal ions (Mn–Cu). Neither of these factors equals three, which might naively be expected as ethene is a two-electron donor while benzene can donate up to six electrons. This clearly reflects that the overlap between the π-cloud of ethene with the metal ion is better than for benzene where the metal is situated close to the center of the ring [130].

The binding energies of the second benzene ligand do not show the same kind of periodic trend as the other series of bond energies. We rationalized this observation by noting that $Mn^+(C_6H_6)_2$ is an 18-electron species. Thus, later metals (Fe–Cu) have weaker binding because the second benzene ligand can no longer donate six electrons effectively. A more complete analysis of these trends also notes that the 18-electron $Mn^+(C_6H_6)_2$ complex is likely to have a singlet spin. Indeed, most of the bis-benzene complexes are likely to have different spin states compared to the atomic metal ion, such that any analysis of the trends in the bond energies must include the energy required to promote the metal to the appropriate electronic state. When this factor is included, we find that the total binding energy of $Mn^+(C_6H_6)_2$ is higher than the other metals, reflecting the stability of the 18-electron complex [121]. Metals to the left of Mn have slightly weaker total binding energies, while those to the right are much weaker.

3.3.3
Alkanes

In addition to conventional ligands like alkenes, benzene, CO, water and ammonia, we have also measured the binding energies of alkanes to a couple of metal ions, Co^+ and Fe^+ [122–126]. These results compare well with other determinations using a different experimental technique [131]. Such thermochemistry is relevant to the alkane activation studies discussed in Sect. 4 because the interaction between the metal ions and the alkanes provides part of the driving force for these reactions. As expected, alkanes are not as good a ligand as more conventional ligands, which have accessible lone pairs of electrons. Nevertheless, the interactions between alkanes and metal ions are appreciable. For instance, the data on methane complexes (Table 2) show that these binding energies are about half the strength of the alkene bond energies. As the alkane gets larger, the strength of this interaction increases because it is determined primarily by the polarizability of the alkane. Thus, $D_0(M^+-C_2H_6)$ are 64 ± 6 [123] and 100 ± 5 [52] kJ mol^{-1} for M=Fe and Co, respectively, while $D_0(M^+-C_3H_8)$ are 75 ± 4 [132] and 129 ± 6 [125] kJ mol^{-1}, respectively.

4
Mechanisms for Alkane Activation

A recurring theme in understanding the activation of molecules at transition metal centers is the orbital interactions. A number of state-specific studies of the reactions of atomic transition metal ions with small molecules, specifically dihydrogen, have indicated the orbital preferences for efficient reactivity [16, 22, 55, 134]. In agreement with simple donor-acceptor theory, these results illustrate that reactions are enhanced by electronic configurations in which there is an empty acceptor orbital on the metal into which the electrons of the bond to be broken are donated. If this acceptor orbital is occupied, there is a repulsive interaction that leads to inefficient reaction either by more direct pathways or by introduction of a barrier to reaction. If this acceptor orbital is empty, the reactants can get sufficiently close so that metal electrons in orbitals having π-symmetry

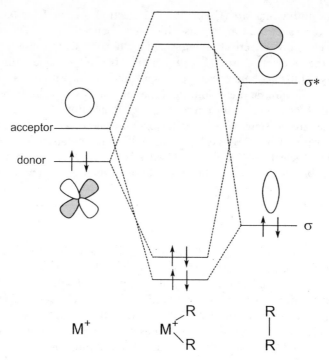

Fig. 4. Molecular orbital interactions for the oxidative addition of a covalent bond to a metal center. Electron occupations for the most efficient situation are shown

can back-donate into the antibonding orbital of the bond to be broken. These considerations are shown schematically in Fig. 4. Both of these acceptor-donor interactions are needed to fully activate the bond.

4.1
Methane

Reactions of transition metal ions with methane can be straightforwardly understood by considering the mechanism shown in Scheme 1. Oxidative addition of a C–H bond to the metal center creates a hydrido methyl transition metal cation intermediate, $HMCH_3^+$. At low energies, the most favorable reaction is dehydrogenation to form $MCH_2^+ + H_2$. The two main mechanisms that can be considered are a four-centered concerted elimination from the $HMCH_3^+$ intermediate and a hydrogen shift to form a $H_2MCH_2^+$ intermediate which then decomposes by reductive elimination of H_2. Both pathways pass through an $(H_2)MCH_2^+$ intermediate (not shown in Scheme 1) in which a hydrogen molecule is electrostatically bound to the MCH_2^+ product.

At high energies, the $HMCH_3^+$ species decomposes more readily by breaking the M–C bond to form $MH^+ + CH_3$. Competing with this is the cleavage of the

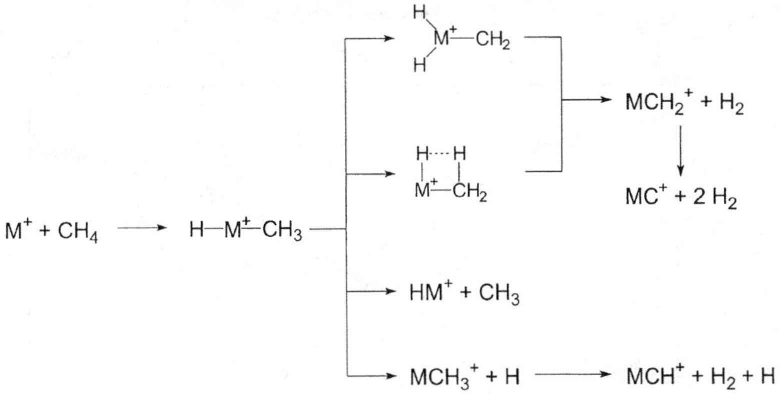

$$\text{M}^+ + \text{CH}_4 \longrightarrow \text{H}-\overset{+}{\text{M}}-\text{CH}_3 \begin{cases} \longrightarrow \underset{\text{H}}{\overset{\text{H}}{\diagdown}}\overset{+}{\text{M}}-\text{CH}_2 \\ \longrightarrow \underset{\text{M}^+-\text{CH}_2}{\overset{\text{H}\cdots\text{H}}{\mid\quad\mid}} \longrightarrow \text{MCH}_2^+ + \text{H}_2 \longrightarrow \text{MC}^+ + 2\,\text{H}_2 \\ \longrightarrow \text{HM}^+ + \text{CH}_3 \\ \longrightarrow \text{MCH}_3^+ + \text{H} \longrightarrow \text{MCH}^+ + \text{H}_2 + \text{H} \end{cases}$$

Scheme 1.

M–H bond to yield MCH_3^++H. These two processes have similar energetics (because the M^+–H and M^+–CH_3 bond energies are fairly similar), but the former reaction is strongly favored. This is because it conserves angular momentum more easily (a result of the relative reduced mass of the MCH_3^++H products being much smaller than those of the M^++CH_4 reactants or the MH^++CH_3 products). These simple bond cleavage reactions involve only loose TSs (because the interaction between the products is an attractive one) while that for dehydrogenation involves more complicated rearrangements and a tight TS. Thus, competition between these channels is easily observed as a suppression of MCH_2^++H_2 formation. At still higher energies, the MCH_2^+ and MCH_3^+ products can decompose by dehydrogenation to form MC^+ and MCH^+, respectively, by M–C bond cleavage to reform M^+, and by H atom loss at still higher energies.

Although Scheme 1 provides a general mechanism for the interaction of transition metals with methane, there is a distinct variation in the reactivities of different metals with methane. The behavior relies on the electronic configurations of the transition metals and on the strength of spin-orbit interactions. These vary the topology of the potential energy surfaces in ways that can be straightforwardly understood. This is discussed in the Sects. 4.1.1–4.1.3.

4.1.1
Early First Row Transition Metal Ions

The transition metal ions found early in the first row, Sc^+, Ti^+, V^+ and Cr^+, all react with methane similarly [35, 135–137], although the endothermicities of the reactions vary from metal to metal. The example of V^+ is shown in Fig. 5. Dehydrogenation to form VCH_2^++H_2 is the lowest energy process and begins at its thermodynamic threshold. This indicates that there are no barriers in excess of the endothermicity for this reaction. It is not known definitively whether dehydrogenation proceeds by the four-center elimination process or by the metal di-

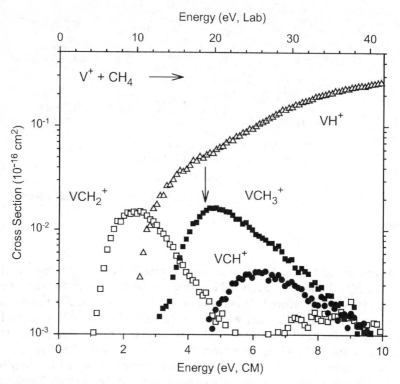

Fig. 5. Cross sections for the reaction of V$^+$ with methane as a function of kinetic energy in the center-of-mass frame (lower x-axis) and laboratory frame (upper x-axis). The arrow indicates $D(H–CH_3)$

hydrido methylidene cation intermediate. However, one imagines that a similar mechanism is followed for the first row metals and theoretical studies have shown that the four-centered TS is operative for Fe$^+$ and Co$^+$, as discussed in the next section. In addition, formation of the dihydride intermediate might be expected to yield a VH$_2$$^+$+CH$_2$ product as well, and no such product is observed.

The cross section for formation of VCH$_2$$^+$ reaches a maximum where the formation of VH$^+$ begins. This shows that the two products must share a common intermediate but formation of the hydride is kinetically favored. The mechanism of Scheme 1 is consistent with these observations. The next product observed is formation of VCH$_3$$^+$. The ratio of the VH$^+$ and VCH$_3$$^+$ product cross sections is consistent with a long-lived intermediate that decomposes into both of these products [135]. The VCH$_3$$^+$ product reaches a maximum near the energy where this product can decompose to V$^+$+CH$_3$, i.e. beginning at $D(H–CH_3)$=4.5 eV. This product also dehydrogenates to form VCH$^+$ and loses a hydrogen atom to give the second feature in the VCH$_2$$^+$ cross section starting near 7 eV.

4.1.2
Late First Row Transition Metal Ions

Among the most detailed results obtained in our laboratories have been for the reactions of Fe^+ and Co^+ with methane [75, 76]. Detailed mechanisms and potential energy surfaces have been elucidated by examining the reactions of the atomic metal ion with methane, the reactions of the metal methylidene cation with H_2 (and D_2), and collisional activation of the metal cation-methane adduct [109, 124]. Results for the reaction of Co^+ with methane (deuterium labeled to enhance mass separation of the products) are shown in Fig. 6. Comparison of these results with those for vanadium (Fig. 5) shows similar behavior with the exception of the dehydrogenation channel to form $CoCD_2^+$. This threshold is well above the thermodynamic threshold for this process (1.5 eV) and appears nearly coincident with that for formation of $CoCD_3^+$.

Examination of the reverse exothermic reaction, $CoCH_2^+ + D_2 \rightarrow Co^+ + CH_2D_2$, confirms that there is a barrier to this reaction lying above the energy of the $CoCH_2^+ + D_2$ species [75]. In addition, the thermoneutral H/D exchange reaction to form $CoCHD^+ + HD$ is observed, but with an apparent threshold lying even

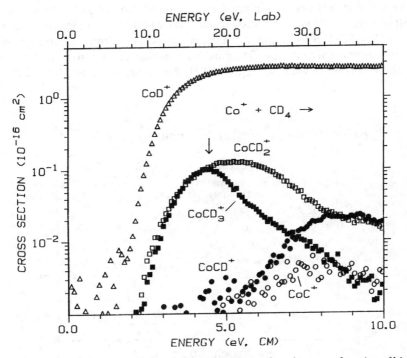

Fig. 6. Cross sections for the reaction of Co^+ with deuterated methane as a function of kinetic energy in the center-of-mass frame (lower x-axis) and laboratory frame (upper x-axis). The arrow indicates $D(D-CD_3)=4.58$ eV. Reprinted with permission from Haynes CL, Chen YM, Armentrout PB (1995) J Phys Chem 99:9110. (1995 American Chemical Society)

higher than that for Co^+ formation. Cobalt hydride (CoH^+ and CoD^+) and cobalt methyl ($CoCH_2D^+$ and $CoCHD_2^+$) cations are also observed beginning at their thermodynamic thresholds.

These observations can be explained using the potential energy surface shown in Fig. 7. The barrier to the $Co^+ + CH_4 \rightarrow CoCH_2^+ + H_2$ reaction and its reverse (and its deuterated analogues) is associated with the four-centered TS shown in Scheme 1. The geometry of this tight TS has been identified by theory [138]. Our analysis of our results for the reverse reaction included detailed modeling using phase space theory (a statistical theory that explicitly includes the conservation of angular momentum). This model reproduces the energy behavior of all the products, including the energy behavior of both Co^+ and $CoCHD^+$ products. These species have the same energy onsets but the latter product is suppressed because it must pass over the tight TS twice. In addition, the best reproduction of our data was achieved by including a TS for the oxidative addition of the C–H bond at the metal, TS2 in Fig. 7. Interestingly, the energy of this TS is essentially the same as that of the $H–M^+–CH_3$ intermediate (estimated from bond additivity), in good agreement with theoretical calculations [113, 138]. Hence, it represents a constraint on the number of available states because of geometric rather than energetic reasons.

Very similar results are obtained for the interaction of Fe^+ with methane [76]; however, there is the additional complexity in this system that the ground state of Fe^+, $^6D(4s^13d^6)$, cannot form the $H–Fe^+–CH_3$ intermediate in a spin-allowed process. This intermediate must have a quartet spin ground state and therefore correlates with $Fe^+(^4F, 3d^7) + CH_4$ reactants, which lie about 0.28 eV

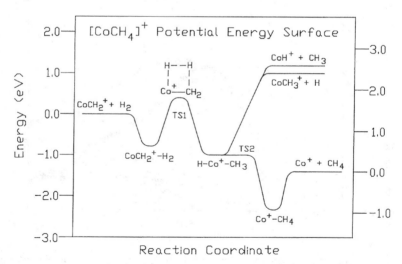

Fig. 7. [$CoCH_4^+$] potential energy surface derived from experimental results. Reprinted with permission from Haynes CL, Chen YM, Armentrout PB (1995) J Phys Chem 99:9110. (1995 American Chemical Society)

above the ground state. State-specific cross sections for reaction of $Fe^+(^4F)$ are very similar to those shown in Fig. 6 for $Co^+(^3F, 3d^8)$, while those for $Fe^+(^6D)$ are smaller by factors of 20 or more and exhibit apparent thresholds well in excess of the thermodynamic limits. This simply illustrates that the ground state no longer reacts primarily by insertion of the metal ion into a C–H bond, but rather by direct abstraction processes. The differences in reactivity are straightforwardly explained using the orbital ideas discussed above in the introduction to Sect. 4.

Reactions of the early metals with methane deserve more detailed experimental and theoretical study. Until that information is available, however, it is interesting to speculate as to why the dehydrogenation channel behaves so differently for early and late transition metal systems. The fundamental difference is the availability of empty valence orbitals for the early metals. The effect that this has can be seen by considering the interaction of MCH_2^+ with H_2. The valence electron configurations of $VCH_2^+(^3B_2)$ and $CoCH_2^+(^3A_2)$ [105, 113] are $(3a_1)^2(1b_1)^2(2b_2)^1(4a_1)^1$ and $(3a_1)^2(1b_1)^2(1a_2)^1(2b_2)^2(4a_1)^2 (5a_1)^1$, respectively, where the $3a_1$ and $1b_1$ orbitals are the σ and π M–C bonding orbitals, respectively; the $1a_2$, $2b_2$, and $4a_1$ orbitals are $3d$-like nonbonding; and the $5a_1$ is $4s$-$3d\sigma$ nonbonding. Based on the orbital ideas discussed above, the key difference in these two orbital populations is the occupation of the $5a_1$ orbital, which acts as the acceptor orbital when H_2 approaches MCH_2^+. Having this orbital occupied leads to the barrier shown in Fig. 7, while having it empty allows a stronger interaction between MCH_2^+ and H_2, allowing the barrier to fall below the $MCH_2^+ + H_2$ asymptote.

4.1.3
Second and Third Row Transition Metal Ions

Studies performed at room temperature [139–144] have shown that the second and third row transition metal ions are more reactive with methane than their first row congeners. The most obvious reflection of this is the dehydrogenation reaction to form MCH_2^+. The reaction $CH_4 \rightarrow CH_2 + H_2$ requires 454 kJ mol^{-1} such that only those metal ions with M^+–CH_2 bond energies greater than this can dehydrogenate methane in an exothermic reaction. Examination of the data in Table 1 shows that the second row metals, Zr^+ and Nb^+, have bond energies slightly below this limit such that these reactions are only slightly endothermic [145, 146]. The observation of an efficient dehydrogenation reaction in room temperature studies [139–142] shows that several third row metal cations (Ta^+, W^+, Os^+, Ir^+ and Pt^+) have bond energies in excess of this limit [139]. Theoretical calculations are consistent with these observations [113, 147].

However, in addition to changes induced by the thermochemistry, it is also found that the late second row metals (Ru^+ and Rh^+) also react much more efficiently with methane than their first row congeners [148, 149]. As shown in Fig. 8, the reactivity of Rh^+ with methane is similar to that observed for the early first row metals. The dehydrogenation reaction begins promptly at its thermo-

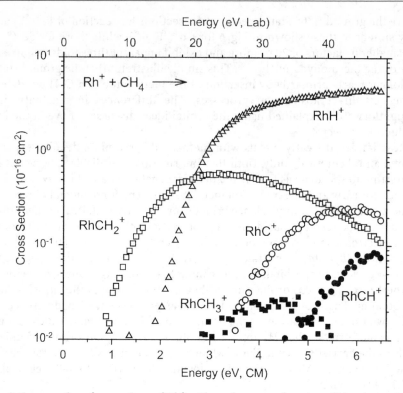

Fig. 8. Cross sections for reactions of Rh$^+$ with methane as a function of kinetic energy in the center-of-mass frame (lower x-axis) and laboratory frame (upper x-axis). Reprinted with permission from Chen YM, Armentrout PB (1995) J Phys Chem 99:10775. (1995 American Chemical Society)

dynamic limit of 0.75 eV. Both RhCH$_2^+$ and RhCH$_3^+$ products readily dehydrogenate to form RhC$^+$ and RhCH$^+$, respectively, again a consequence of the much stronger π-bonds for the second row metals (see Sect. 3.2.1).

We have rationalized this behavior with the help of theoretical calculations [113, 150]. These show that the triplet surface, which correlates Rh$^+$(^{3}F)+CH$_4$ reactants with excited RhCH$_2^+$(^{3}A$_2$)+H$_2$ products, is similar to that shown for the Co$^+$ system in Fig. 6. However, the ground state of RhCH$_2^+$ is ^{1}A$_1$ with a $(3a_1)^2(1b_1)^2(1a_2)^2(2b_2)^2(4a_1)^2$ $(5a_1)^0$ electron configuration [105, 113, 150, 151]. As discussed in the previous section, the empty $5a_1$ orbital means that this species can interact strongly with H$_2$, thereby reducing the barrier along the singlet surface associated with the four-centered TS of Scheme 1. However, this dehydrogenation reaction is spin-forbidden. As the threshold for RhCH$_2^+$ formation in Fig. 8 corresponds to formation of the ground ^{1}A$_1$ state, there must be effective spin-orbit coupling interactions that allow a facile transition from the triplet to the singlet state in this reaction.

The third row congener of Co^+ and Rh^+, Ir^+, is known to react at room temperature with methane to form $IrCH_2^+$ [139]. Although detailed experimental results at higher kinetic energies are not yet available, it is worth mentioning the theoretical results of Perry concerning this process [113]. He found that the lowest energy pathway for dehydrogenation now involves a stable $H_2IrCH_2^+$ dihydride intermediate rather than the four-centered TS used in the Co and Rh systems. Spin-orbit interactions are sufficiently strong that the reaction of $Ir^+(^5F)$ inserts into the C–H bond to form a triplet $H–Ir^+–CH_3$ species which rearranges by α-H migration to a singlet $H_2IrCH_2^+$ intermediate (a dihydride carbene that is the global minimum on the potential energy surface). This species reductively eliminates H_2 while crossing back to a triplet $(H_2)IrCH_2^+$ electrostatically bound intermediate before final H_2 expulsion to yield $IrCH_2^+(^3A_2)$. Perry attributes the strongly bound $H_2IrCH_2^+$ intermediate to the effectiveness of sd hybridization for the third row metal and the accessibility of different electronic configurations. Some experimental evidence for the stability of the $H_2MCH_2^+$ species for third row metals comes from the observation of MH_2^+ products in reactions with methane at higher kinetic energies [152].

4.2
Ethane

In many respects, the reactions of transition metal cations with ethane are similar to those observed for methane. Figures 9 and 10 show results for the first and second row transition metal congeners, Co^+ and Rh^+. At high energies, C–H and now C–C bond cleavage reactions form products like $MH^++C_2H_5$, $MH+C_2H_5^+$ (which dehydrogenates at higher energies to form $C_2H_3^+$), and MCH_3^+ (which dehydrogenates at higher energies to yield MCH^+). Methane elimination to form MCH_2^+, reaction (7), parallels the dehydrogenation reaction with methane to form the same product ion. This product can dehydrogenate at higher energies to form MC^+.

$$M^++C_2H_6 \rightarrow MCH_2^++CH_4 \tag{7}$$

The ethane system does show distinct differences from the reactions with methane because the energy required to dehydrogenate ethane requires much less energy than that required to dehydrogenate methane, 129 vs 454 kJ mol^{-1}. However, M^+–ethene bond energies are also much less than M^+–CH_2 bond energies. Examination of the data in Table 2 shows that reaction (8) is exothermic for all first row transition metal cations except V^+, Cr^+ and Mn^+.

$$M^++C_2H_6 \rightarrow M^+(C_2H_4)+H_2 \tag{8}$$

Experimentally, it has been found that only Sc^+ and Ti^+ undergo reaction (8) at thermal energies [135, 153]. Thus, there must be a barrier in excess of the reactant energy for $M^+=Fe^+–Cu^+$. This is clearly shown in Fig. 9 for M=Co. Although the data is not complete for second row metals, it is known that Y^+ [135], Zr^+ [145], Nb^+ [146], Ru^+ [148] and Rh^+ [154] do undergo reaction (8) at ther-

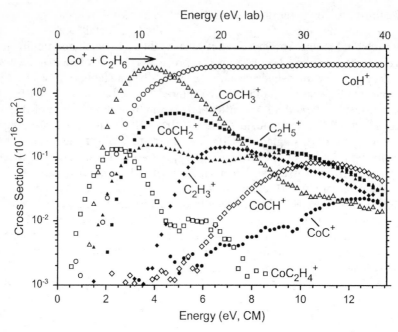

Fig. 9. Cross sections for reactions of Co^+ with ethane as a function of kinetic energy in the center-of-mass frame (lower x-axis) and laboratory frame (upper x-axis). Reprinted with permission from Haynes CL, Fisher ER, Armentrout PB (1996) J Am Chem Soc 118:3269. (1996 American Chemical Society)

mal energies, while Mo^+ [146], Pd^+ [155] and Ag^+ [156] do not. This is evident in Fig. 10 for M=Rh, where the $Rh^+(C_2H_4)$ product clearly shows no barrier to formation. At higher energies, this product further dehydrogenates to form the $Rh^+(C_2H_2)$ product ion. Therefore, similar to the reactions with methane, the first and second row congeners exhibit different reactivity, presumably for the same fundamental reasons described above.

The details of the potential energy surfaces for interaction of Co^+ with ethane have recently been examined experimentally [126]. The stability of the electro-statically bound $Co^+(C_2H_6)$ complex intermediate was determined as 100 ± 5 kJ mol^{-1}, in reasonable agreement with values from equilibrium measurements [157] and theory [158, 159]. In addition, the kinetic energy dependence of both the forward and reverse of reactions (7) and (8) were studied. For reaction (7), a barrier 28 ± 9 kJ mol^{-1} above the products was determined, which directly parallels the results for dehydrogenation of methane by Co^+ where the barrier height was 34 ± 8 kJ mol^{-1}. This barrier was assigned to a four-centered TS directly analogous to that shown in Scheme 1 (where a methyl group replaces one of the hydrogens). For reaction (8), a barrier of 31 ± 12 kJ mol^{-1} above the reactants was

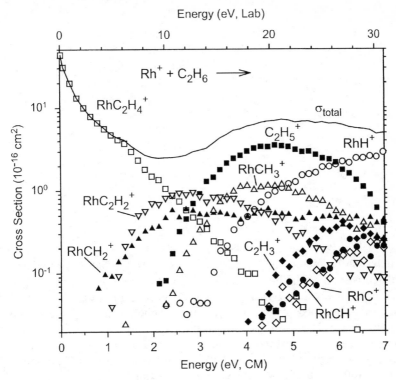

Fig. 10. Cross sections for reactions of Rh$^+$ with ethane as a function of kinetic energy in the center-of-mass frame (lower x-axis) and laboratory frame (upper x-axis). Reprinted with permission from Chen YM, Armentrout PB (1995) J Am Chem Soc 117:9291. (1995 American Chemical Society)

determined. Studies of the reactions of ethane with Rh$^+$ [154], the second row congener of Co$^+$, find no barriers for either reaction (7) or (8).

There are several plausible mechanisms that have been considered for reaction (8) and these are shown in Scheme 2. Both mechanisms involve initial oxidative addition of C–H bonds to the metal center. For many years, it was generally thought that the initial step was formation of a hydrido ethyl metal cation intermediate **1**. Then, a β-H from the ethyl group would migrate to the metal center forming the dihydrido metal ethene cation **4**, which seems reasonable because such β-H transfers are facile in solution. However, calculations by Perry [113] and Holthausen and Koch [159] for M=Co indicate that intermediate **1** is not a minimum on the potential energy surface (see Sect. 3.2.2). They found that the concerted multicenter TS **3** shown in Scheme 2 is the lowest energy pathway for the dehydrogenation reaction induced by Co$^+$. They were unable to find any evidence for a stable H$_2$Co$^+$(C$_2$H$_4$) species **4**, a result that can be understood using the same orbital concepts developed above. Formation of two covalent bonds

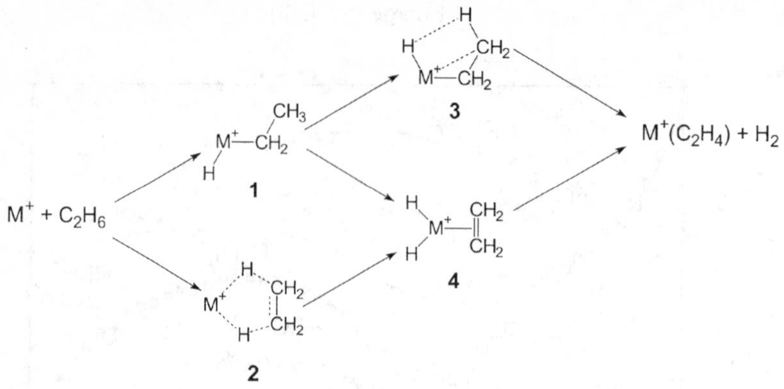

$$M^+ + C_2H_6 \longrightarrow \begin{matrix} \text{(1)} & M^+\text{-CH}_2 / \text{CH}_3, \text{H} \\ \text{(2)} & M^+ \cdots \text{CH}_2 / \text{CH}_2, \text{H}, \text{H} \end{matrix} \longrightarrow \begin{matrix} \text{(3)} & M^+\text{-CH}_2 / \text{CH}_2, \text{H}, \text{H} \\ \text{(4)} & M^+ \cdots \text{CH}_2 \| \text{CH}_2, \text{H}, \text{H} \end{matrix} \longrightarrow M^+(C_2H_4) + H_2$$

Scheme 2.

to late transition metal ions is achieved by using $4s$-$3d\sigma$ hybrid orbitals on the metal (consistent with the correlation with s^1d^n promotion noted above in Sect. 3.2.2). For Co, this gives a valence electron configuration of $\sigma^2\sigma^2d^6$ where the six d electrons occupy four nonbonding d orbitals as the fifth d orbital is involved in the covalent σ-bonds. This gives a triplet ground state for this species with no empty valence orbitals on the metal that are available to form a strong dative bond with the ethene ligand. Similar considerations should hold for $M^+=$ Fe^+-Cu^+ [160], the first row metal cations that exhibit barriers to the dehydrogenation reaction. Alternatively, one can think about the bonding by examining the reverse of the dehydrogenation, i.e. oxidative addition of H_2 to $M^+(C_2H_4)$. A strong M^+-C_2H_4 bond is formed by donation of the ethene π-electrons into an empty s orbital on the metal. As the s orbital is now occupied, the metal no longer has a good acceptor orbital needed to efficiently activate the H_2 bond. Perry calculates that oxidative addition of H_2 to $Co^+(C_2H_4)$ is a higher energy process than oxidative addition to Co^+ by about 70 kJ mol^{-1}. Hence, oxidative addition of H_2 to $Co^+(C_2H_4)$ occurs in a multicenter reaction, TS **3**, although this species still lies higher in energy than the reactants, explaining the barrier observed.

In contrast, Perry finds that intermediate **4** is stable for the second row metal ion Rh^+. This is because s-d hybridization is much more efficient, the d orbitals can be used to form strong covalent bonds without hybridization (see Sect. 3.2.1), and better spin-orbit coupling allows facile transfer to the singlet surface that is the ground state of $H_2Rh^+(C_2H_4)$. The singlet spin coupling allows the six nonbonding electrons to be placed in three orbitals such that there is an empty orbital available on RhH_2^+ to form a strong dative bond with ethene. Perry calculates that oxidative addition of H_2 to $Rh^+(C_2H_4)$ is a lower energy process than oxidative addition to Rh^+ by about 70 kJ mol^{-1}, exactly the opposite result compared with Co^+. Interestingly, although intermediate **4** is stable for M= Rh, Perry calculates that **1** lies higher than the reactant energy and therefore proposes that the exothermic dehydrogenation of ethane by Rh^+ occurs by the synchronous oxidative addition of two C–H bonds in TS **2**. The complexity of these reactions is then indicated by the results of his calculations for Ir^+. Here, the lowest energy pathway is formation of **1** followed by **4**, i.e. the classic mechanism long believed to be operative for the lighter congeners. Thus, Co^+ dehydrogenates ethane by a **1–3** mechanism; Rh^+ uses **2–4**; while Ir^+ goes by **1–4**. At this time, there is no experimental work that can definitively verify the existence of these disparate pathways.

4.3
Propane

More attention has been paid to the reactions of atomic transition metal ions with propane than to any other hydrocarbon system. This is because propane is the smallest alkane for which exothermic reactions are observed at thermal energies, making them amenable to study by many experimental techniques. Both dehydrogenation, reaction (9), and methane elimination, reaction (10), are observed with branching ratios that are sensitive functions of the metal. This is illustrated in Table 3.

$$M^+ + C_3H_8 \rightarrow M^+(C_3H_6) + H_2 \tag{9}$$

$$M^+ + C_3H_8 \rightarrow M^+(C_2H_4) + CH_4 \tag{10}$$

For many metal cations, these reactions are exothermic and exhibit no barriers in excess of the reactant energy; however, the efficiency varies considerably with metal identity as shown in Table 3. At higher energies, metal ions react with propane along pathways that are similar to those observed with ethane, i.e. a host of C–H and C–C bond cleavage products are formed in endothermic reactions.

Table 3. Reaction efficiencies and branching ratios for transition metal cation reactions with propane at thermal energies

		% Product	
M	$\sigma_{tot}/\sigma_{col}$[a]	$M^+(C_3H_6)$	$M^+(C_2H_4)$
Ti[b]	0.18	98	2
V[b]	0.005	100	
Cr[c]	0		
Mn[d]	0		
Fe[e]	0.08	26	74
Co[f]	0.13	77	23
Ni[e]	0.13	20	80
Cu[g]	0		
Zr[h]	1.00	89*	11*

Table 3. Reaction efficiencies and branching ratios for transition metal cation reactions with propane at thermal energies (continued)

M	$\sigma_{tot}/\sigma_{col}$[a]	% Product	
		$M^+(C_3H_6)$	$M^+(C_2H_4)$
Nb[i]	1.00	98*	2*
Mo[i]	0		
Ru[j]	1.00	100	
Rh[k]	1.00	100	
Pd[l]	0.03	25	75
Ag[m]	0		

[a] σ_{col} is the collision cross section calculated as $\pi e(2\alpha/E)^{1/2}$. [b][164]. [c]Fisher ER, Armentrout PB (1992) J Am Chem Soc 114:2039. [d]This behavior is inferred from studies in Sunderlin LS, Armentrout PB (1990) J Phys Chem 94:3589. [e][161]. [f][74]. [g]Georgiadis R, Fisher ER, Armentrout PB (1989) J Am Chem Soc 111:4251. [h][145]. [i][146]. [j][148]. [k][154]. [l][155]. [m][156]. *These products are also observed to dehydrogenate again at thermal energies.

There have been a number of detailed experimental and theoretical studies that have focused on determining the mechanisms for the propane reactions [73, 74, 113, 159–164]. For the most part, the mechanisms involved directly parallel the discussion for ethane above. Hence, only particularly interesting aspects of these studies will be discussed in this section.

4.3.1
Late First Row Transition Metal Ions

The earliest comprehensive study of the mechanisms of propane reactions was a collaborative effort between our laboratory and those of van Koppen, Bowers, and Beauchamp [73, 74]. We examined the kinetic energy dependence of the reactions of Co^+ with several deuterated propanes: d_0, 2-d_1, 2,2-d_2, 1,1,1-d_3, 1,1,1,3,3,3-d_6, and d_8. Kinetic energy release distributions (KERDs) of the $Co^+(C_3H_6)$ and $Co^+(C_2H_4)$ product ions were also measured. Using phase space theory, we showed that the inefficiency of this reaction (only 13% of the ions form products at thermal energies, Table 3) can quantitatively be explained by a barrier to reaction lying *below* the asymptotic energy of the reactants (modeled as 11±3 kJ mol^{-1} below). Although energetically accessible, $Co^+(C_3H_8)$ intermediates formed with high angular momentum cannot traverse this barrier while still conserving angular momentum. [Because the TS is tight, its moment of inertia, I, is smaller than that for the loose (or orbiting) TS evolving back to $Co^+ + C_3H_8$ reactants. Thus, for a given rotational angular momentum, L, the rotational energy, $E_{rot} = L^2/2I$, is larger for the tight TS than the loose TS. If too much energy is tied up in rotational motion, there may not be enough for motion along the reaction coordinate. This situation is shown in Fig. 11.] We also

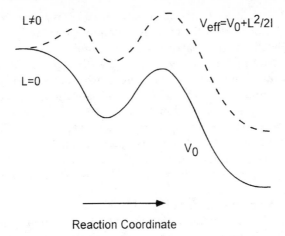

$$V_{eff} = V_0 + L^2/2I$$

Reaction Coordinate

Fig. 11. Schematic potential energy surfaces illustrating a tight transition state on the way to products. The lower surface (*full line*) shows the potential energy surface in the absence of angular momentum (L=0) while the upper surface (*dashed line*) indicates the effective potential energy including the contributions of E_{rot} (see text)

found that deuterium substitution decreases the reaction efficiency by an amount consistent with the increase in the potential barrier associated with a C–H vs a C–D bond energy, i.e. a zero point energy difference of about 5 kJ mol^{-1}. This demonstrates that the rate-limiting TS involves making or breaking a C–H bond. These studies also demonstrated that this tight TS affects the methane elimination channel and, hence, we concluded that a single rate-limiting TS was involved in both reactions (9) and (10). At the time, it was natural to designate this tight TS as the oxidative addition of a C–H bond to the metal center, i.e. the TS leading to an intermediate analogous to **1** in Scheme 2 (except there is a propyl rather than an ethyl group attached to the metal). Subsequent work of a similar quantitative nature was performed for the analogous Fe$^+$ and Ni$^+$ reactions with propane, with similar conclusions drawn [161]. Rate-limiting barriers lying 7±3 and 10±3 kJ mol^{-1} below the respective reactant asymptotes were suggested in this work.

Since these initial studies, theory [113, 159, 160, 162] has decisively eliminated intermediates analogous to **1** from consideration for reactions of Fe$^+$, Co$^+$ and Ni$^+$, as discussed in Sect. 3.2.2. Instead, theory indicates that the rate-limiting TS is a multicenter process analogous to **3** in Scheme 2. Our detailed study of reactions (9) and (10) in the forward and reverse directions for Co$^+$ [125] and very recent time-dependent studies of the Ni$^+$/propane system [163] are both successfully interpreted in terms of such multicenter TSs. Our collision-induced dissociation studies of M$^+$(C$_3$H$_8$) intermediates, where M=Fe and Co [125, 132], provide direct evidence that there is a barrier to H$_2$ and CH$_4$ elimination that lies below the energies of the M$^+$+C$_3$H$_8$ asymptotes, in agreement with our previous findings [73, 74]. In contrast to the conclusion that there is a single rate-limiting TS lying 11 ± 3 kJ mol^{-1} below the reactant asymptote, however, this work indicates that the H$_2$ elimination barrier is 28 ± 4 kJ mol^{-1} below this asymptote, while CH$_4$ elimination occurs via two processes with barriers 8 ± 8 kJ mol^{-1} below and 5 ± 4 kJ mol^{-1} above for M=Co. These two processes are assigned as those proceeding from initial C–H and C–C bond activation, respectively. As the conclusions of both the earlier and more recent studies are heavily model dependent, we believe that these results are not in direct conflict. Similarly, the work of Weisshaar and co-workers [163] is interpreted in terms of *two* rate limiting multicenter TSs both lying about 10 kJ mol^{-1} below the Ni$^+$+C$_3$H$_8$ reactants, in very good agreement with our earlier value of 10 ± 3 kJ mol^{-1} but for a single TS [74].

Another key observation in our original studies was bimodal KERDs for H$_2$ elimination [73, 74]. Both a statistically behaved feature and one giving high kinetic energies were observed. Further, the affect of deuterium substitution on these two regions is distinctly different. These isotope effects led us to conclude that the statistical component originated from activation of the primary C–H bonds, while the high energy component came from activation of the secondary C–H bonds and evolution through a multicenter TS. However, if multicenter TSs are the lowest energy pathways, then there is no longer a simple explanation for the bimodal KERDs. In our more recent work [125], we suggested several possibilities for the bimodality, although no clear cut explanation could be realized. Weisshaar and co-workers [162] have speculated that electronically excited states of the reactant metal ions might be responsible for the high energy component of the KERDs. Additional experimental work is needed to further elucidate these particular observations and reconcile them with theory, although it is also possible that theory has not identified all the reaction pathways that are energetically accessible in these systems.

4.3.2
Early First and Second Row Transition Metal Ions

As discussed in Sect. 4.2, the intermediates analogous to **4** in Scheme 2 are unstable for late first row transition metal ions, but not second and third row metal ions. Hence, it is not surprising to see that the reactivities of Ru$^+$ and Rh$^+$ are much higher than their first row congeners, Fe$^+$ and Co$^+$ (Table 3). However, these second row metal ions react at thermal energies exclusively by dehydrogenation, reaction (9). Reaction (10) is exothermic but exhibits barriers. Clearly, dehydrogenation can occur by a mechanism directly analogous to that for ethane, **2–4** in Scheme 2, but it is difficult to specify a likely pathway for the methane elimination channel. Several possibilities were discussed in our work on the Rh$^+$ system [154].

In addition, the discussion above points to the absence of unoccupied orbitals in species like **1** or MH$_2^+$ as the key reason that these molecules cannot form a strong dative bond with alkenes. Hence, intermediates analogous to **4** may be energetically feasible for early first row transition metal ions. This issue has not be examined theoretically at this time, but we have examined the reactions of Ti$^+$ and V$^+$ [164] using the comprehensive approach noted above for Fe$^+$, Co$^+$ and Ni$^+$ (deuterium-labeling studies and KERDs). Similar to the second row transition metal ions, Ti$^+$ and V$^+$ react with propane at thermal energies by reaction (9) almost exclusively (Table 3). The dehydrogenation reaction can be explained by processes analogous to the 1–3 or 1–4 sequences in Scheme 2; however, a multicenter process does not provide a reasonable interpretation of the isotope effects observed for the methane elimination channel. Here, reductive elimination of CH$_4$ from an intermediate analogous to 4 (where CH$_3$ replaces one of the H atom ligands) is believed to be rate limiting. A particularly interesting aspect of this study involves the potential energy surfaces believed to be involved in the reactions of Ti$^+$ and V$^+$. We argue that reactions (9) and (10) for Ti$^+$ involve coupling from the quartet spin surfaces of the reactants to doublet spin surfaces of the intermediates and products. Reactions with V$^+$ involve coupling from the quintet spin surfaces associated with ground state reactants to triplet spin surfaces for the intermediates and then back to quintet spin surfaces for the products.

4.4
Effect of Ancillary Ligands

The periodic trends in the reactivities of atomic transition metal cations with alkanes demonstrate that the reaction efficiency and pathways are strongly influenced by the electron populations in the orbitals. It should also be possible to systematically influence the reactivity by ligation of the transition metal center. Such ligands need not participate directly in the bond activation chemistry but can alter the orbital populations on the metal. We have recently conducted studies of the effects of two such ancillary ligands, H$_2$O and CO, on the reactivity of Fe$^+$ with D$_2$ [165], CH$_4$ and C$_2$H$_6$ [166], and C$_3$H$_8$ [167]. [It should be noted that CO can conceivably participate directly in the reactions, for instance, by forming aldehydes or ketones. There is some evidence for this in the propane system, but this represents less than 5% of the observed reactivity.] Figure 12 shows this for the example of reactions with propane, specifically, reactions (9) and (10).

Qualitatively, we find that both ligands suppress the overall reactivity of the metal ion. This is not surprising as donation of electron density from the ligands (which are both σ-donors) to the metal should partially occupy the σ-acceptor orbital on the metal which must be empty for efficient activation of covalent bonds (see Sect. 4). However, CO suppresses the reactivity much more than H$_2$O, which we believe is an effect associated with the π-symmetry orbitals. H$_2$O is a π-donating ligand and hence can enhance the ability of the metal to back-donate electron density to the antibonding orbitals of the covalent bond to be broken. In contrast, CO is a π-accepting ligand that will suppress such back-donation.

Probably the most interesting aspect of the results shown in Fig. 12 is the large change in branching ratios. Table 3 indicates that atomic iron cations form three

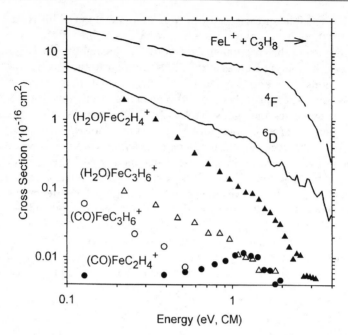

Fig. 12. Cross sections for the exothermic reactions of ligated iron cations with propane as a function of kinetic energy in the center-of-mass frame (lower x-axis). *Triangles* and *circles* show results for $Fe^+(H_2O)$ and $Fe^+(CO)$ reactants, respectively. Dehydrogenation and demethanation processes are shown by *open* and *closed symbols*, respectively. Total cross sections for reactions (9) and (10) with $Fe^+(^6D)$ and $Fe^+(^4F)$ are indicated by *solid* and *dashed lines*, respectively. Reprinted with permission from Armentrout PB, Tjelta BL (1997) Organometallics 16:5372. (1997 American Chemical Society)

times as much $Fe^+(C_2H_4)$ as $Fe^+(C_3H_6)$. This ratio is independent of the kinetic energy and of the electronic state of Fe^+. For $Fe^+(H_2O)$, the C–C bond activation channel is still favored, but by a factor of 20–30 at low kinetic energies. In contrast, $Fe^+(CO)$ favors C–H bond activation, the analogue of reaction (9), by at least an order of magnitude and the analogue of reaction (10) now exhibits a barrier even though the overall reaction is undoubtedly exothermic. Note that the strong differences in how the ligands affect the C–C vs C–H bond activation channels suggests that these pathways have independent rate-limiting TSs, in agreement with our most recent conclusions regarding these potential energy surfaces (see Sect. 4.3). We again attribute these strong differences to the different π-character of the ligands and the influence that this has on the rate-limiting TSs analogous to **1** in Scheme 2. Clearly, theoretical calculations would be useful to further elucidate this interesting effect.

A related consequence of the ligation can be seen in the thermochemistry of FeH^+ and $FeCH_3^+$ with and without ligands attached. The bond energies (Table 4) show that the water ligand does not grossly affect the covalent bonds between iron and H or CH_3, while the CO ligand drastically reduces these bond

Table 4. Bond energies of ligated Fe^+ in kJ mol^{-1} [166]

L	Fe^+-L	$(H_2O)Fe^+$-L	$(CO)Fe^+$-L
H	205(6)	215(15)	127(8)
CH_3	229(5)	190(10)	132(10)

energies. This is believed to be a consequence of the different electronic config-urations of $Fe^+(H_2O)$ and $Fe^+(CO)$. Calculations indicate that the ground state of $Fe^+(CO)$ is $^4\Sigma^-$ $(3d\sigma^1 3d\pi^4 3d\delta^2)$ [168], while $Fe^+(H_2O)$ has a $^6A_1$ $(4s^1 3d\sigma^1 3d\pi^2 3d\delta^3)$ ground state with a low-lying $^4A_1$ $(3d\sigma^1 3d\pi^2 3d\delta^4)$ excited state [169, 170]. Note the π-orbital populations properly indicate that CO is a π-acceptor and H_2O is a π-donor. As found in Sect. 3.2, first row transition metal ions form strong covalent bonds using the $4s$ orbital. Therefore, the electronic configuration of $Fe^+(H_2O)$ is suited to formation of strong bonds with H and CH_3, while $Fe^+(CO)$ must either form a weaker bond using the $3d\sigma$ orbital or promote an electron to the $4s$ orbital, thereby destabilizing the Fe^+–CO bond.

4.5
Effect of an Oxo Ligand

In the previous section, we considered recent studies addressing the affects of ligands that do not actively participate in the reaction chemistry. Interesting chemistry can also be found where the ligand is an integral component of the processes observed. A technologically important example is the oxo (atomic ox-ygen) ligand. Many recent studies have considered the oxidation of hydrocar-bons by transition metal oxide cations and there is an excellent review of this topic [171]. One of the interesting features of our early work on such oxidation systems was the recognition that spin conservation was an important aspect of these processes. For instance, we found that MO^+, where M=Sc, Ti and V, would oxidize H_2 in an endothermic reaction to form $M^+ + H_2O$; however, the thresh-olds measured for these reactions indicated that excited states of M^+ (especially those where spin was conserved) were preferentially formed [172]. This propen-sity for spin-conservation is also evident in state-specific studies of the reverse reaction, $M^+ + H_2O \rightarrow MO^+ + H_2$ [173, 174].

The early transition metal oxide ions are not efficient oxidizers because the MO^+ BDEs are very strong, 550–700 kJ mol^{-1} [52, 175]. More interesting oxidiz-ers are the later metal oxide cations where the BDEs range from 160–360 kJ mol^{-1} [52, 175]. These diatomic metal oxide cations will exothermically oxidize methane to methanol as the process, $CH_4 + O \rightarrow CH_3OH$, releases 371 kJ mol^{-1}. As prototypical studies, we have examined the oxidation of H_2 and CH_4 by FeO^+ and CoO^+ and their reverse reactions in some detail [176–178]. These results can be fruitfully compared with that discussed above for the bare metal cation, $M^+ + CH_4$, and with $MCH_2^+ + CH_4$, the reverse of reaction (7).

4.5.1
Reaction of CoO⁺ with Methane

Results for the reaction of CoO^+ with methane are shown in Fig. 13. As for Co^+ (Fig. 6), the dominant reaction is hydrogen atom transfer. This reaction is much less endothermic (23 ± 10 kJ mol^{-1}) than for the bare metal ion, which indicates that the product is a cobalt hydroxide ion as an O–H bond is much stronger than a Co–H bond. At higher energies, methyl transfer to form $CoOCH_3^+$ is also observed, in analogy with the $CoCH_3^+$ product observed for the bare metal ion reactions. The CoH^+ product observed at higher energies in this system is formed in the most efficient decomposition pathway of the $CoOCH_3^+$ species, a cobalt methoxide ion. This dissociation channel, loss of formaldehyde, has been explicitly studied in the analogous case of $FeOCH_3^+$ [179]. $CoOCH_3^+$ also decomposes by losing a hydrogen atom at higher energies to yield $CoOCH_2^+$.

Clearly, the primary difference between the reactivity displayed in Figs. 6 and 13 is the observation of the Co^+ product in the CoO^+ system. No analogous re-

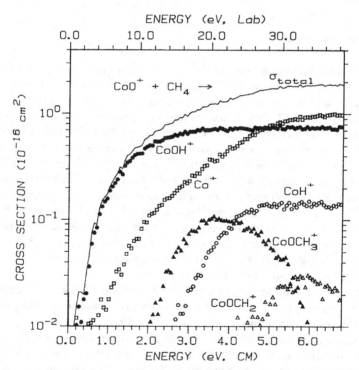

Fig. 13. Cross sections for the reaction of CoO^+ with methane as a function of kinetic energy in the center-of-mass frame (lower x-axis) and laboratory frame (upper x-axis). Reprinted with permission from Chen YM, Clemmer DE, Armentrout PB (1994) J Am Chem Soc 116:7815. (1994 American Chemical Society)

action is possible in the Co^+/CH_4 system. This product can be formed in reactions (11) and (12).

$$CoO^+ + CH_4 \rightarrow Co^+ + CH_3OH \tag{11}$$

$$CoO^+ + CH_4 \rightarrow Co^+ + O + CH_4 \tag{12}$$

As the bond energy of CoO^+ is 314 ± 5 kJ mol^{-1} (3.25 ± 0.05 eV) [52, 175], the simple collision-induced dissociation process (12) cannot occur until kinetic energies above this value. Hence, the formation of Co^+ observed at lower energies must result from the oxidation of methane to methanol. The complex shape of the Co^+ cross section clearly indicates that both reactions (11) and (12) occur. Oddly, reaction (11) is exothermic by 58 ± 5 kJ mol^{-1}, yet our results clearly indicate that there is a barrier to the process. It should be noted that ICR results of this same system find that reaction (11) does occur at thermal energies, albeit with a small efficiency of 0.005 ± 0.005 [180]. This corresponds to a cross section that should be easily observed on the scale of Fig. 13. The quantitative discrepancy between these studies has an unclear origin, especially given that similar results for the analogous reaction of FeO^+ with H_2 and methane are in good accord [178]. Nevertheless, both studies agree that reaction (11) is inefficient even though exothermic.

The observation of a barrier to this reaction is similar to our results for reaction of $CoCH_2^+ + CH_4$ [126]. This latter system is directly analogous to the reaction of $CoCH_2^+ + H_2$ discussed in detail in Sect. 4.1.2 and the origin of the barrier is the same in both. As noted above, $CoCH_2^+$ has a $CoCH_2^+(^3A_2)$ ground state with a $(3a_1)^2(1b_1)^2(1a_2)^1(2b_2)^2(4a_1)^2(5a_1)^1$ valence electron configuration. The barrier is largely a result of the repulsive interactions between the occupied $5a_1$ orbital and the electrons in the covalent bond being activated. For CoO^+, theory [181] finds that the ground state is $^5\Delta$ with a $(2\sigma)^2(1\pi)^4(1\delta)^3(2\pi)^2(3\sigma)^1$ valence electron configuration where the 2σ and 1π are the M–O bonding orbitals (parallel with the $3a_1$ and $1b_1$ orbitals of MCH_2^+); the 1δ and 2π (like the $1a_2, 2b_2$, and $4a_1$) orbitals are $3d$-like nonbonding; and the 3σ (like the $5a_1$) orbital is $4s$-$3d\sigma$ nonbonding. In essence, the valence configurations of CoO^+ and $CoCH_2^+$ are very similar, both having an occupied $4s$-$3d\sigma$ nonbonding orbital that leads to the barrier observed in the reactivity. Indeed, the barrier observed for the reaction, $CoCH_2^+ + CH_4 \rightarrow Co^+ + C_2H_6$, is measured as 27 ± 10 kJ mol^{-1}, while that for the analogous reaction (11) is 54 ± 8 kJ mol^{-1}. One difference between the two systems is that the former reaction is spin-allowed, while reaction (11) is spin-forbidden from ground state reactants to ground state products. Hence, triplet surfaces correlating with excited states of CoO^+ (which are calculated to lie about 1.0–1.4 eV above the ground state [181]) must play a role in this oxidation process. The most influential of these will be for a $^3\Sigma^-$ state having a $(2\sigma)^2(1\pi)^4(1\delta)^4(2\pi)^2(3\sigma)^0$ valence electron configuration in which the 3σ orbital is empty.

4.5.2
Reaction of FeO⁺ with Methane

Evidence for such a curve crossing comes from studies of the reaction of FeO^+ with CH_4 which has been studied in detail [178, 182]. Our work finds that the observed reactivity parallels that shown in Fig. 13 except that cross sections for formation of $FeOH^++CH_3$ and Fe^++CH_3OH exhibit distinct exothermic features in addition to endothermic features like those seen for reactions of CoO^+. (Very similar differences are observed for the reactions of FeO^+ and CoO^+ with D_2 [176, 177].) The kinetic energy dependence of these exothermic features is consistent with spin-forbidden pathways. Hence, the endothermic features in the reactions of FeO^+ (and presumably CoO^+) result from spin-allowed pathways that must surmount a barrier to the overall process. As before, this barrier is attributed to the four-centered TS associated with oxidative addition of a C–H (or H–H) bond across the metal oxide bond. The exothermic features in the FeO^+ system correspond to spin-forbidden pathways where the barriers are below the reactant asymptote. Apparently, in the CoO^+ system, the barrier for the low-spin state must lie above the reactant energy asymptote. Detailed theoretical calculations of the iron systems confirm these qualitative aspects of the FeO^+ system [181, 183, 184].

4.5.3
Reaction of Other Transition Metal Oxide Cations with Methane

On the basis of the results discussed above, it would clearly be of interest to examine the reactions of other transition metal oxide cations with methane. One might anticipate that second and third row metals may oxidize methane much more efficiently because spin-orbit coupling will permit much more efficient reactions along multiple potential energy surfaces. At present, however, only a few systems have been examined experimentally. For example, it has been observed that NiO^+ undergoes a reaction in analogy to reaction 11 with modest efficiency (20% at thermal energies) [171]. PtO^+, the third row congener, reacts with methane more efficiently (on every collision), but the primary product is $PtCH_2^++H_2O$, and formation of methanol constitutes only 25% of the products [3]. MnO^+, like FeO^+, abstracts a hydrogen atom from methane to form $MnOH^++CH_3$, but this is the only process observed at thermal energies [185]. OsO^+ reacts with methane by dehydrogenation to form $O=Os=CH_2^+$ [141]. Note that the reactions of both third row metal cations are driven by formation of strong $M=CH_2$ bonds, consistent with the thermochemistry discussed above.

5
Conclusions

Clearly, gas-phase studies of organometallic chemistry are a rich source of quantitative information. As this review demonstrates, it is certainly unmatched in its ability to provide specific metal–ligand bond energies, details of the mechanisms for reactions of organic molecules at transition metal centers, and insight into the effects of spin and electronic configuration on reactivity. Because nearly all species studied are unsaturated, reactive species, periodic trends in this quantitative information are readily obtained. By examining these periodic trends, the information can be reassembled and placed in a context that permits the specific quantitative information to be generalized. It is the trends in the thermodynamics and chemistry that can be translated from the gas phase to real condensed phase catalytic systems.

In parallel with the progress made in understanding the characteristics needed for efficient and selective reactions at atomic transition metal centers, the isolation of the gas phase also allows investigations of the systematic effects of selective ligation and solvation on reaction energetics and mechanisms. In essence, this permits reactive intermediates present in condensed phase systems to be approached and mimicked in the gas-phase studies. The observation that different ligands affect both the reactivity and the *selectivity* of a chemical process is a fascinating one that has implications for developing new catalytic reagents. Clearly, this is a fertile area for continued studies that will further bridge our knowledge of gas-phase and condensed phase organometallic chemistry.

References

1. Kappes MM, Staley RH (1981) J Am Chem Soc 103:1286
2. Schröder D, Fiedler A, Ryan MR, Schwarz H (1994) J Phys Chem 98:68
3. Wesendrup R, Schröder D, Schwarz H (1994) Angew Chem 106:1232; Angew Chem Int Ed Engl 33:1174
4. Shi Y, Ervin KM (1998) J Chem Phys 108:1757
5. Armentrout PB, Hales DA, Lian L (1994) In: Duncan MA (ed) Advances in metal and semiconductor clusters, vol 2. JAI, Greenwich, p 1
6. Armentrout PB (1996) In: Russo N, Salahub DR (eds) Metal-ligand interactions – structure and reactivity. Kluwer, Dordrecht, p 23
7. Armentrout PB; Griffin JB; Conceição J (1999) In: Cheuv GN, Lakhno VD, Nefedov AP (eds) Progress in physics of clusters. World Scientific, Singapore
8. Allison J, Freas RB, Ridge DP (1979) J Am Chem Soc 101:1332
9. Tonkyn R, Ronan M, Weisshaar JC (1988) J Phys Chem 92:92
10. For a recent review, see MacMillan DK, Gross ML (1989) In: Russell DH (ed) Gas phase inorganic chemistry. Plenum, New York, p 369
11. Garstang RH (1962) Mon Not R Astron Soc 124:321, personal communication
12. Freas RB, Ridge DP (1980) J Am Chem Soc 102:7129
13. Halle LF, Armentrout PB, Beauchamp JL (1981) J Am Chem Soc 103:962
14. Reents WD, Strobel F, Freas RB, Wronka J, Ridge DP (1985) J Phys Chem 89:5666
15. Strobel F, Ridge DP (1989) J Phys Chem 93:3635
16. Elkind JL, Armentrout PB (1985) J Phys Chem 89:5626
17. Elkind JL, Armentrout PB (1986) J Chem Phys 84:4862
18. Elkind JL, Armentrout PB (1986) J Phys Chem 90:5736
19. Elkind JL, Armentrout PB (1986) J Am Chem Soc 108:2765
20. Elkind JL, Armentrout PB (1986) J Phys Chem 90:6576
21. Elkind JL, Armentrout PB (1987) J Chem Phys 86:1868
22. Elkind JL, Armentrout PB (1988) Int J Mass Spectrom Ion Processes 83:259

23. Kemper PR, Bowers MT (1991) J Phys Chem 95:5134
24. van Koppen PAM, Kemper PR, Bowers MT (1992) J Am Chem Soc 114:10941
25. Schultz RH, Crellin KC, Armentrout PB (1991) J Am Chem Soc 113:8590
26. Kang H, Beauchamp JL (1985) J Phys Chem 89:3364
27. Loh SK, Fisher ER, Lian L, Schultz RH, Armentrout PB (1989) J Phys Chem 93:3159
29. Tonkyn R, Weisshaar JC (1986) J Am Chem Soc 108:7128
30. Buckner SW, Freiser BS (1987) J Am Chem Soc 109:1247
31. Huang Y, Freiser BS (1988) J Am Chem Soc 110:4434
32. Saha MN (1920) Phil Mag Series 6 40:472
33. Langmuir I, Taylor JB (1933) Phys Rev 34:423
34. Wilson RG, Brewer GR (1973) Ion beams with applications to ion implantation. Wiley-Interscience, New York
35. Sunderlin LS, Armentrout PB (1988) J Phys Chem 92:1209
36. Sanders L, Sappy AD, Weisshaar J (1986) J Chem Phys 85:6952
37. Sanders L, Hanton S, Weisshaar JC (1987) J Phys Chem 91:5145
38. Armentrout, PB (1990) Annu Rev Phys Chem 41:313
39. Schultz RH, Armentrout PB (1991) Int J Mass Spectrom Ion Processes 107:29
40. Armentrout PB (1995) Acc Chem Res 28:430
41. Honma K, Sunderlin LS, Armentrout PB (1993) J Chem Phys 99:1623
42. Fisher ER, Kickel BL, Armentrout PB (1993) J Phys Chem 97:10204
43. Dalleska NF, Honma K, Armentrout PB (1993) J Am Chem Soc 115:12125
44. Lehmen T A, Bursey M M (1976) Ion cyclotron resonance spectrometry. Wiley, New York
45. Weiting RD, Staley RH, Beauchamp JL (1975) J Am Chem Soc 97:924, 5920
46. Ervin K, Armentrout PB (1985) J Chem Phys 83:166
47. Teloy E, Gerlich D (1974) Chem Phys 4:417
48. Gerlich D (1992) In: Ng C-Y, Baer M (eds), State-selected and state-to-state ion-molecule reaction dynamics. Part 1: experiment. Wiley, New York, p 1
49. Forbes RA, Lech LM, Freiser BS (1987) Int J Mass Spectrom Ion Processes 77:107
50. Hop CECA, McMahon TB (1991) J Phys Chem 95:10582
51. Beyer M, Bondybey VE (1997) Rapid Comm Mass Spectrom 11:1588
52. Armentrout PB, Kickel BL (1996) In: Freiser BS (ed) Organometallic ion chemistry. Kluwer, Dordrecht, p 1
53. Freiser BS (ed) (1996) Organometallic ion chemistry. Kluwer, Dordrecht
54. Armentrout PB, Clemmer DE (1992) In: Simoes JAM (ed) Energetics of organometallic species. Kluwer, Dordrecht, p 321
55. Armentrout PB (1989) In: Russell DH (ed) Gas phase inorganic chemistry. Plenum, New York, p 1
56. Armentrout PB (1990) In: Davies JA, Watson PL, Liebman JF, Greenberg A (eds) Selective hydrocarbon activation: principles and progress. VCH, New York, p 467
57. Aristov N, Armentrout PB (1986) J Phys Chem 90:5135
58. Loh SK, Hales DA, Lian L, Armentrout PB (1989) J Chem Phys 90:5466
59. Hales DA, Armentrout PB (1990) J Cluster Science 1:127
60. Hales DA, Lian L, Armentrout PB (1990) Int J Mass Spectrom Ion Processes 102:269
61. Robinson PJ, Holbrook KA (1972) Unimolecular reactions. Wiley, London
62. Khan FA, Clemmer DE, Schultz RH, Armentrout PB (1993) J Phys Chem 97:7978
63. Rodgers MT, Ervin KM, Armentrout PB (1998) J Chem Phys 106:4499
64. Talrose VL, Vinogradov PS, Larin IK (1979) In: Bowers MT (ed) Gas phase ion chemistry, vol 1. Academic, New York, p 305
65. Armentrout PB (1992) In: Adams NG, Babcock LM (eds) Advances in gas phase ion chemistry, vol 1. JAI, Greenwich, p 83
66. Georgiadis R, Armentrout PB (1986) J Am Chem Soc 108:2119
67. Ervin KM, Armentrout PB (1986) J Chem Phys 84:6738

68. Ervin KM, Armentrout PB (1987) J Chem Phys 86:2659
69. Weber ME, Elkind JL, Armentrout PB (1986) J Chem Phys 84:1521
70. Elkind JL, Armentrout PB (1984) J Phys Chem 88:5454
71. Boo BH, Armentrout PB (1987) J Am Chem Soc 109:3549
72. Armentrout PB, Halle LF, Beauchamp JL (1982) J Chem Phys 76:2449
73. van Koppen PAM, Brodbelt-Lustig J, Bowers MT, Dearden DV, Beauchamp JL, Fisher ER, Armentrout PB (1990) J Am Chem Soc 112:5663
74. van Koppen PAM, Brodbelt-Lustig J, Bowers MT, Dearden DV, Beauchamp JL, Fisher ER, Armentrout PB (1991) J Am Chem Soc 113:2359
75. Haynes CL, Chen YM, Armentrout PB (1995) J Phys Chem 99:9110
76. Haynes CL, Chen YM, Armentrout PB (1996) J Phys Chem 100:111
77. Armentrout PB, Simons J (1992) J Am Chem Soc 114:8627
78. Holland PM, Castleman AW (1980) J Am Chem Soc 102:6174
79. Holland PM, Castleman AW (1982) J Chem Phys 76:4195
80. Peterson KI, Holland PM, Keesee RG, Lee N, Mark TD, Castleman AW (1981) Surf Sci 106:136
81. Buckner SW, Freiser BS (1988) Polyhedron 7:1583
82. Magnera TF, David DE, Michl J (1989) J Am Chem Soc 111:4100
83. Marinelli PJ, Squires RR (1989) J Am Chem Soc 111:4101
84. van Koppen PAM, Bowers MT, Beauchamp JL, Dearden DV (1990) ACS Symp Ser 428:34
85. Armentrout PB, Sunderlin LS (1992) In: Dedieu A (ed) Transition metal hydrides. VCH, New York, p 1
86. Schilling JB, Goddard WA, Beauchamp JL (1986) J Am Chem Soc 108:582
87. Schilling JB, Goddard WA, Beauchamp JL (1987) J Am Chem Soc 109:5565
88. Schilling JB, Goddard WA, Beauchamp JL (1987) J Phys Chem 91:5616
89. Rappe AK, Upton TH (1986) J Chem Phys 85:4400
90. Pettersson LGM, Bauschlicher CW, Langhoff SR, Partridge H (1987) J Chem Phys 87:481
91. Bauschlicher CW, Langhoff SR, Partridge H, Barnes LA (1989) J Chem Phys 91:2399
92. Schilling JB, Goddard WA, Beauchamp JL (1987) J Am Chem Soc 109:5573
93. Rosi M, Bauschlicher CW (1989) J Chem Phys 90:7264
94. Rosi M, Bauschlicher CW (1990) J Chem Phys 92:1876
95. See discussion in Aristov N, Armentrout PB (1986) J Am Chem Soc 108:1806
96. Chesnavich WJ, Bowers MT (1979) J Phys Chem 83:900
97. Sunderlin LS, Aristov N, Armentrout PB (1987) J Am Chem Soc 109:78
98. Armentrout PB, Beauchamp JL (1981) J Chem Phys 74:2819
99. Armentrout PB, Beauchamp JL (1981) J Am Chem Soc 103:784
100. Chantry PJ (1971) J Chem Phys 55:2746
101. Lifshitz C, Wu RLC, Tiernan TO, Terwilliger DT (1978) J Chem Phys 68:247
102. Armentrout PB (1990) ACS Symp Ser 428:18
103. Armentrout PB, Georgiadis R (1988) Polyhedron 7:1573
104. Carter EA, Goddard WA (1988) J Phys Chem 92:2757
105. Bauschlicher CW, Partridge H, Sheehy JA, Langhoff SR, Rosi M (1992) J Phys Chem 96:6969
106. Sunderlin LS, Armentrout PB (1990) Organometallics 9:1248
107. Sievers MR, Armentrout PB unpublished work
108. Halle LF, Crowe WE, Armentrout PB, Beauchamp JL (1984) Organometallics 3:1694
109. Haynes CL, Armentrout PB, Perry JK, Goddard WA (1995) J Phys Chem 99:6340
110. Rosi M, Bauschlicher CW, Langhoff SR, Partridge H (1990) J Phys Chem 94:8656
111. Blomberg MRA, Siegbahn PEM, Svensson M, Wennerberg J (1992)) In: Simoes JAM (ed) Energetics of organometallic species. Kluwer, Dordrecht, p 387
112. Hendrickx M, Ceulemans M, Gong K, Vanquickenborne L (1997) J Phys Chem 101:2465
113. Perry JK (1994) PhD thesis, Caltech

114. Hendrickx M, Ceulemans M, Vanquickenborne L (1996) Chem Phys Lett 257:8
115. Goebel S, Haynes CL, Khan FA, Armentrout PB (1995) J Am Chem Soc 117:6994
116. Sievers MR, Armentrout PB (1995) J Phys Chem 99:8135
117. Khan FA, Steele DL, Armentrout PB (1995) J Phys Chem 99:7819
118. Meyer F, Chen YM, Armentrout PB (1995) J Am Chem Soc 117:4071
119. Meyer F, Armentrout PB (1996) Mol Phys 88:187
120. Sievers MR, Jarvis LM, Armentrout PB (1998) J Am Chem Soc 120:1891
121. Meyer F, Khan FA, Armentrout PB (1995) J Am Chem Soc 117:9740
122. Schultz RH, Armentrout PB (1991) J Am Chem Soc 113:729
123. Schultz RH, Armentrout PB (1992) J Phys Chem 96:1662
124. Schultz RH, Armentrout PB (1993) J Phys Chem 97:596
125. Haynes CL, Fisher ER, Armentrout PB (1996) J Phys Chem 100:18300
126. Haynes CL, Fisher ER, Armentrout PB (1996) J Am Chem Soc 118:3269
127. Dalleska NF, Honma K, Sunderlin LS, Armentrout PB (1994) J Am Chem Soc 116:3519
128. Walter D, Armentrout PB (1998) J Am Chem Soc 120:3176
129. Sodupe M, Bauschlicher CW, Langhoff SR, Partridge H (1992) J Phys Chem 96:2118
130. Bauschlicher CW, Partridge H, Langhoff SR (1992) J Phys Chem 96:3273
131. Kemper PR, Bushnell J, van Koppen P, Bowers MT (1993) J Phys Chem 97:1810
132. Schultz RH, Armentrout PB (1991) J Am Chem Soc 113:729
134. Elkind JL, Armentrout PB (1987) J Phys Chem 91:2037
135. Sunderlin LS, Armentrout PB (1989) J Am Chem Soc 111:3845
136. Aristov N, Armentrout PB (1987) J Phys Chem 91:6178
137. Georgiadis R, Armentrout PB (1988) J Phys Chem 92:7067
138. Musaev DG, Morokuma K, Koga N, Nguyen KA, Gordon MS, Cundari TR (1993) J Phys Chem 97:11435
139. Irikura KK, Beauchamp JL (1991) J Phys Chem 95:8344
140. Irikura KK, Beauchamp JL (1991) J Am Chem Soc 113:2769
141. Irikura KK, Beauchamp JL (1989) J Am Chem Soc 111:75
142. Buckner SW, MacMahon TJ, Byrd GD, Freiser BS (1989) Inorg Chem 28:3511
143. Ranasinghe YA, MacMahon TJ, Freiser BS (1991) J Phys Chem 95:7721
144. Buckner SW, Freiser BS (1987) J Am Chem Soc 109:1247
145. Sievers MR, Steele D, Armentrout PB unpublished work
146. Sievers MR, Armentrout PB unpublished work
147. Irikura KK, Goddard WA (1994) J Am Chem Soc 116:8733
148. Armentrout PB, Chen YM (1999) J Am Soc Mass Spectrom in press
149. Chen YM, Armentrout PB (1995) J Phys Chem 99:10775
150. Musaev DG, Koga N, Morokuma K (1993) J Phys Chem 97:4064
151. Russo N (1992) In: Salahub DR, Russo N (eds) Metal-ligand interactions: from atoms, to clusters, to surfaces. Kluwer, Dordrecht, p 341
152. Rue C, Armentrout PB unpublished work
153. Sunderlin LS, Armentrout PB (1989) Int J Mass Spectrom Ion Processes 94:149
154. Chen YM, Armentrout PB (1995) J Am Chem Soc 117:9291
155. Chen YM, Sievers MR, Armentrout PB (1997) Int J Mass Spectrom Ion Processes 167/168:195
156. Chen YM, Armentrout PB (1995) J Phys Chem 99:11424
157. Kemper PR, Bushnell J, van Koppen P, Bowers MT (1993) J Phys Chem 97:1810
158. Perry JK, Ohanessian G, Goddard WA (1993) J Phys Chem 97:5238
159. Holthausen MC, Koch W (1996) J Am Chem Soc 118:9932
160. Holthausen MC, Fiedler A, Schwarz H, Koch W (1996) J Phys Chem 100:6236
161. van Koppen PAM, Bowers MT, Fisher ER, Armentrout PB (1994) J Am Chem Soc 116:3780
162. Yi SS, Blomberg MRA, Siegbahn PEM, Weisshaar JC (1998) J Phys Chem 102:395
163. Noll RJ, Yi SS, Weisshaar JC (1998) J Phys Chem 102:386

164. van Koppen PAM, Bowers MT, Haynes CL, Armentrout PB (1998) J Am Chem Soc 120:5704
165. Tjelta BL, Armentrout PB (1995) J Am Chem Soc 117:5531
166. Tjelta BL, Armentrout PB (1996) J Am Chem Soc 118:9652
167. Armentrout PB, Tjelta BL (1997) Organometallics 16:5372
168. Barnes LA, Rosi M, Bauschlicher CW (1990) J Chem Phys 93:609
169. Rosi M, Bauschlicher CW (1989) J Chem Phys 90:7264
170. Rosi M, Bauschlicher CW (1990) J Chem Phys 92:1876
171. Schröder D, Schwarz H (1995) Angew Chem 107:2126; Angew Chem Int Ed Engl 34:1973
172. Clemmer DE, Aristov N, Armentrout PB (1993) J Phys Chem 97:544
173. Clemmer DE, Chen YM, Aristov N, Armentrout PB (1994) J Phys Chem 98:7538
174. Chen YM, Clemmer DE, Armentrout PB (1994) J Phys Chem 98:11490
175. Fisher ER, Elkind JL, Clemmer DE, Georgiadis R, Loh SK, Aristov N, Sunderlin LS, Armentrout PB (1990) J Chem Phys 93:2676
176. Clemmer DE, Chen YM, Khan FA, Armentrout PB (1994) J Phys Chem 98:6522
177. Chen YM, Clemmer DE, Armentrout PB (1994) J Am Chem Soc 116:7815
178. Schröder, D, Schwarz H, Clemmer DE, Chen YM, Armentrout PB, Baranov VI, Bohme DK (1997) Int J Mass Spectrom Ion Processes 161:175
179. Fiedler A, Schröder D, Schwarz H, Tjelta BL, Armentrout PB (1996) J Am Chem Soc 118:5047
180. Ryan MF, Fiedler A, Schröder D, Schwarz H (1994) Organometallics 13:4072
181. Fiedler A, Schröder D, Shaik S, Schwarz H (1994) J Am Chem Soc 116:10734
182. Schröder D, Fiedler A, Hrušák J, Schwarz H (1992) J Am Chem Soc 114:1215
183. Danovich D, Shaik S (1997) J Am Chem Soc 119:1773
184. Yoshizawa K, Shiota, Y, Yamabe T (1997) Chem Eur 3:1160
185. Ryan MF, Fiedler A, Schröder D, Schwarz H (1995) J Am Chem Soc 117:2033

Static and Dynamic Structures of Organometallic Molecules and Crystals

Dario Braga*, Fabrizia Grepioni
D. Braga (e-mail: dbraga@ciam.unibo.it)
Dipartimento di Chimica G. Ciamician, Università di Bologna, Via Selmi 2, I-40126 Bologna, Italy
F. Grepioni (e-mail: grepioni@ssmain.uniss.it)
Dipartimento di Chimica, Università di Sassari, Via Vienna 2, I-07100 Sassari, Italy

This chapter focuses on the structures of organometallic molecules and ions in the solid state as determined by crystallographic methods. The first part is devoted to a discussion of the relationship between static and dynamic structure of flexible organometallic molecules in the solid state, while the second part concentrates on solid state transformations and on the relationship between molecular and crystal dynamics. Leading references to entry points in the various aspects are provided. The aim is to discuss some critical aspects of organometallic structural work highlighting some of the most common traps and problems (both practical and conceptual) that are encountered in the field. Directions of possible future development of the discipline in the booming fields of supramolecular and materials chemistry are also discussed.

Keywords: Organometallics, Solid state, Dynamic behavior, Intermolecular interactions, Polymorphism, Phase transitions

Topics in Organometallic Chemistry, Vol. 4
Volume Editors: J.M Brown and P. Hofmann
© Springer-Verlag Berlin Heidelberg 1999

1
Introduction

Traditionally, chemistry has been concerned with atoms and with bonds between atoms and we are all used to teaching students that a chemical reaction is
the process in which bonds between atoms are broken and formed on going
from reactants to products. Only in recent times has the focus of chemistry
moved from molecules to molecular aggregates and an increasing number of
chemists, and organometallic chemists among them, are now concerned with
extramolecular bonding [1]. This shift of interest has also implied a change in
the energy scale of the reactivity process to control. While covalent bond breaking and forming requires a large amount of energy, the energy of non-covalent
interactions, such as those holding together molecules in the solid state or controlling molecular recognition and supramolecular assembly, is much smaller.
For instance, the enthalpy of sublimation of a molecular solid is of the order of
tens of $kJ \cdot mol^{-1}$, whereas the bonding enthalpy of a C–C or H–H bond is of the
order of hundreds of $kJ \cdot mol^{-1}$ [2].

The "molecular chemistry" approach is still dominant in structural organometallic chemistry, a young field of research when compared with organic solid
state chemistry. The consequence is a quite natural tendency to focus on the
characteristics and properties of the individual molecular entity extracted from
its environment, this attitude being strengthened by the limitations of theoretical back-up techniques to model and rationalize molecular structures of molecules containing one or more heavy transition metal atoms [3]. Theoretical
methods cannot yet deal at a satisfactory level of sophistication with the complexity of supramolecular aggregates containing metal atoms, let alone crystalline materials, while results of high accuracy can be obtained on isolated, gas-
phase-like molecules.

We shall address some less conventional aspects of organometallic structural
chemistry. The first part of the contribution will be devoted to a discussion of the
relationship between the static and dynamic structure of flexible organometallic
molecules in the solid state while the second part will concentrate on solid state
transformations and the relationship between molecular and crystal dynamics.
We will close by discussing problems associated with the use of organometallic
solids in the booming fields of supramolecular and materials chemistry [4].

In spite of the "confinement" to molecular dimensions, the methods of choice
for a comprehensive analysis and thorough appreciation of the structures of reactants and products has been, and still is, chiefly crystallographic in nature, i.e.
based on the investigation of solids. The apparent conceptual contradiction
should be appreciated: we learn most about individual molecules from situations in which they interact most with other molecules (see Fig. 1).

Because of the extraordinary power, portability, and reproducibility of X-ray
based experimental results we tend to forget that the method needs crystals, i.e.
well-organized, ordered solids (and grown to appreciable size!). In order to obtain measurable diffraction intensities, solidity is not sufficient, the collective

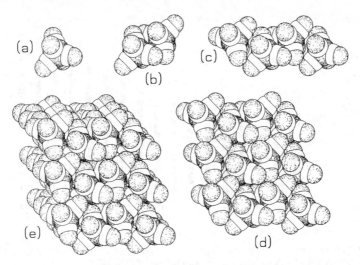

Fig. 1. A schematic representation of the 'aufbau' process leading from molecules to crystals: a molecule→a molecular aggregate→a crystal

contribution of an extremely large number of identical molecules symmetrically distributed through the space of the crystal is a necessary prerequisite. This is the reason why crystallographic experiments only afford *average information* on the aggregated state [5].

As a working definition we shall treat organometallic compounds as those in which the carbon atoms of organic groups are bound to metal atoms via two-electron covalent σ-bonds or via π-interactions of unsaturated groups with metal orbitals of appropriate orientation and symmetry [6]. The organic residue bound to the metal centers is often a molecule itself, that is to say a stable and isolable chemical entity. We have taken the liberty of expanding slightly this definition to encompass, for the sake of the following discussion, also compounds containing "inorganic" carbon, such as binary metal carbonyls, or carrying organic ligands though in the absence of direct M–C bonding. Clearly, the combination of ligands and number of metal atoms has led to an extraordinary level of structural variability in organometallic chemistry.

It is interesting to observe (see Fig. 2) that in 1997 the number of organometallic compounds in the Cambridge Structural Database (CSD) overtook that of organic compounds (87,000 vs. 85,000) [7]. It is easy to predict that in the coming years the diffusion of area detectors will cause the number of crystallographically determined structures to rise dramatically. This will represent per se a new problem and a substantial scientific challenge as it will become quite impossible to approach the data deposited in the CSD (and other databases, of course) without powerful expert systems for the interrogation and assessment of the statistical significance of an extremely large and rapidly expanding set of data [8].

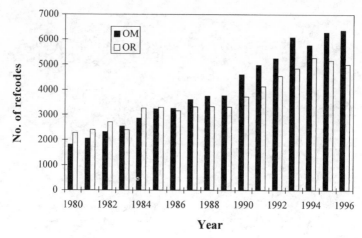

Fig. 2. Comparison of the increase in number of organic (*empty bars*) and organometallic (*shaded bars*) structures deposited in the CSD from 1980 to 1996

Structural non-rigidity is one of the distinctive characteristics of organometallic molecules because of the delocalized nature of the bonding between the metal center(s) and some ligands (e.g. unsaturated π-systems such as aromatic rings, alkenes, alkynes, etc.) and the availability of *almost* isoenergetic, though geometrically different, bonding modes for the same ligand (CO, phosphines, arsines, NO^+, CN^-, etc.). Furthermore, many organometallic molecules exist in different isomeric forms that interconvert via low-energy processes (viz. reorientation, diffusion, scrambling, fluxionality) both in the gas phase and in the condensed state [9]. It is essential to take structural non-rigidity into account when studying organometallic solids with diffraction methods [10]. The relationship between information on static and dynamic aspects as obtained from diffraction experiments will be discussed in the following section.

2
Static and Dynamic Structures of Organometallic Molecules in the Solid State

Single crystal X-ray crystallography is the method of choice for the determination of solid state structures. While the advent of area detector techniques and powerful X-ray radiation sources has enlarged the experimental limits (e.g. molecular and/or unit cell size, number of data to collect and manipulate [11]) the use of variable temperature techniques, combined with spectroscopic solid state methods (such as solid state NMR techniques, e.g. CPMAS [12]) now permits the investigation (or reinvestigation) of the temperature dependence of solid state molecular and crystal structures and the study of the phase transitional behavior associated with solid state dynamics.

Although this is a very well-known fact to all chemists (and some readers may even regard it as irritatingly naive), the final structural information is almost invariably taken by the user as concerning an individual molecule, generally depicted as "hanging in the emptiness", with the accompanying plethora of bond lengths and angles, molecular dimensions, intramolecular non-bonding contacts, torsions, etc. The sceptical reader may browse any chemistry journal and count the number of computer graphics pictures of molecules or ions with relative captions describing "the molecular structure of ...".

IR, NMR, UV and other structural techniques also provide average structural information depending on the time scale of the experiment (see below). The basic difference, however, is that the processing of experimental X-ray diffraction data affords something very specific and useful, namely atomic coordinates (though sometimes in unfriendly crystallographic references systems). The availability of coordinates means that three-dimensional representations of the molecular structure can be drawn as shown in Fig. 3. Nothing is more satisfying to the palate of the synthetic chemist than the possibility of actually looking at the result of the difficult, sometimes costly, and often frustrating work of synthesis, purification, isolation, and crystallization of reaction products. In the course of the process that leads from synthesis to characterization of the products the limitations of the crystallographic answer are often forgotten, and the structure determined by means of single crystal (or powder) diffraction is taken as the ultimate and conclusive result of a whole chemical process. This is still the most common attitude in the field of organometallic chemistry, where the characterization of structurally complex new species relies and depends on the (rapid) structural determination of the reaction products. This approach is not adequate when the structure is not rigid but flexible, i.e. adaptable to some extent to the needs of close packing.

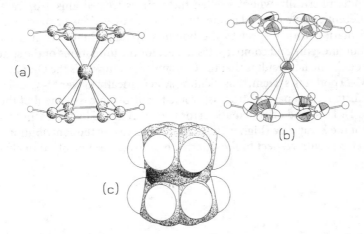

Fig. 3. Three alternative representations of a molecular structure obtained from crystallographic coordinates : "balls and sticks", ORTEP-like with a representation of anisotropic displacement parameters, and "space-filling" models using van der Waals radii

2.1
Single Crystal X-ray Diffraction: Some Limitations

There are many sources of errors in routine single crystal X-ray diffraction experiments. These are discussed extremely well in a number of text books and publications and there is no need to go into too much detail here [13]. There are some aspects, however, that impact more specifically on the reliability of the structural information on organometallic crystals.

The spherical atom approximation is particularly relevant in organometallic crystallography. It is common practice in the processing of crystallographic data to use atomic scattering factors for spherical atoms, i.e. for atoms "at rest in the emptiness" [14]. This situation is, of course, very far from that of atoms in molecules, where outer shell electrons take part in bonding systems. Asphericity affects particularly those atoms with few electrons taking part in multiple bonding and/or participating in very asymmetric bonding interactions. This is commonplace in coordination chemistry where, for instance, metal-bound 10-electron ligands, such as CO, CN^- and NO^+, have on one hand fairly diffuse metal–ligand bonding interactions (via the well-known σ-π donation back donation bonding system) while on the other electron density is accumulated between the two light atoms. Under these conditions the spherical atom treatment may lead to gross errors in the positioning of the electron density centroid of the middle atom in the M–X–Y bonding system [15]. The combination of atomic displacement with severe deviation from sphericity causes the so-called "sliding effect" commonly observed on passing from isotropic to anisotropic treatment of metal-bound CO ligands. The effect is more appreciable in the terminal bonding mode than when the ligand bridges two or three metal atoms. When the atom is treated isotropically (i.e. allowed the same displacement in all directions) the electron density is spread out spherically, whereas when the atom is treated anisotropically (i.e. with different extents of displacement in the three directions of space and with free orientation with respect to the bonding system) the electron density is spread out unevenly and compensates in part for the limitations of the spherical atom model. The net result is that the C-atom "slides" towards the O-atom along the M–C–O bonding system, thus yielding systematically longer M–C and shorter C–O distances on passing from isotropic to anisotropic treatment of the data (see Fig. 4). This kind of systematic error may be reduced by increasing the resolution of the X-ray data (higher $2\theta_{max}$), i.e. by increasing the contribution of the core electrons with respect to the external ones which are involved in chemical bonds.

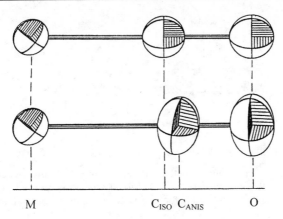

M   C$_{ISO}$ C$_{ANIS}$ O

Fig. 4. Representation of the "sliding effect" observed on passing from isotropic (*top*) to anisotropic (*bottom*) refinement of diffraction data for the C- and O-atoms of terminally bound CO ligands

2.2
Dynamical Processes at Molecular Level

Structural flexibility is related to atomic displacement in the solid state. Organometallic molecules have delocalized bonding interactions which are not common in organic solids and are almost totally absent in inorganic systems. Ligands bound in delocalized fashion possess motional freedom in the solid state even though the whole molecular object may be static. As a matter of fact ligands such as benzene, hexamethylbenzene, thiophene and cyclopentadienyl ligands, undergo reorientational jumping motions in the solid state with low or very low activation energies or potential barriers provided that the bonding to metal centers is delocalized. High motional freedom is reflected in extensive in-plane displacement of the ligands (see Fig. 5). The opposite is not true: i.e. extensive displacement may result from static disorder due to imperfect overlap of the electron density averaged out via Bragg's law in the extremely large number of unit cells that form the crystal (see below) [16]. The study of the relationship between static and dynamic disorder may afford useful insight into bonding features both at the molecular and intermolecular level. It is, however, necessary to deal with the problem of time scale first.

2.3
The Timescale in Diffraction Experiments

A good understanding of the timescale problem becomes essential when dynamic phenomena are investigated by a combination of spectroscopic and diffraction techniques. It is well known that each type of structural technique de-

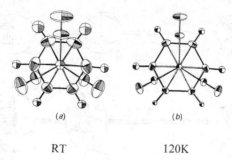

(a)   (b)

RT   120K

Fig. 5. Temperature dependence of the in-plane motion of the benzene ligand in $(C_6H_6)Mo(CO)_3$: note how the extensive librational motion is reduced on passing from room temperature to 120 K indicating that the anisotropic displacement parameters reflect true atomic motion

livers information about a molecular structure averaged over a characteristic timescale, thus giving a measure of the lifetime of the species under investigation [17]. In general terms the definition of timescale depends on the characteristics of both the technique employed and the process under investigation. In absorption spectroscopy the timescale is related to the frequency span of the spectral range and to the lifetime of an absorbing group. Short lifetimes give broadened spectral lines so that structurally non-equivalent groups that undergo site exchange at a rapid enough rate cannot be distinguished because the corresponding spectral lines are not resolved. NMR spectroscopy is best suited to tackle dynamic phenomena because the timescale may vary within a very large interval (10^{-1}–10^{-9} s). NMR gives a time-average of atomic positions in the form of a coalesced spectrum when the exchange rate is near or higher than the separation (in frequency units) between individual resonances. In vibrational spectroscopy (IR, Raman), the combination of a higher radiation frequency and a wider spectral range produces a much shorter timescale (ca. 10^{-13} s). As virtually all molecular rearrangements take much longer than 10^{-13} s, IR spectroscopy gives essentially information on the ground-state structure of molecules which may be non-rigid on the NMR timescale. Coalescence is not observed in IR experiments except in very few cases.

Contrary to NMR and IR, the time required for the X-ray radiation to interact with the electrons (ca. 10^{-18} s) has no relevance for a diffraction experiment. Clearly, if this were the true timescale of the diffraction experiment, the structure would necessarily correspond to a molecule in its ground-state form. Since radiation is diffracted and not absorbed the interaction time does not affect how the experimental technique "sees" the molecules. Diffraction intensities are a time-average of all possible atomic displacements averaged again over the entire crystal, and thus contains information on all atomic motions that take place in the crystal. On this premise it is clear that the timescale of a diffraction experiment corresponds to the entire duration of the data collection [18]. It should be kept in mind that atomic displacements (often erroneously called "thermal parameters") may be either due to motion or to misplacement of atoms or atomic groups through space, viz. crystallographic studies at only one temperature cannot discriminate between static and dynamic disorder.

2.4
Atomic Displacements

Variable temperature measurements may reveal how the molecules and the crystals behave with temperature, hence may allow a distinction to be made between those parameters that change with temperature and those which do not. Anisotropic displacement parameters may thus provide a clear indication of preferential motion in particular directions. However, a discrimination between genuine atomic motion and displacement is indeed possible only if the temperature dependence of the anisotropic displacement parameters is known and/or if complementary information (mainly of spectroscopic source) on the dynamic nature of the phenomenon under investigation is available.

Ferrocene provides a benchmark example [19]. Dunitz and Seiler have discovered from the analysis of the anisotropic displacement parameters of the C-atoms collected on monoclinic ferrocene at room temperature and at 173 K that the staggered conformation of the two cyclopentadienyl rings is the result of static disorder. The staggered structure of monoclinic ferrocene switches to a triclinic crystal containing almost eclipsed molecules via a phase transition at 163.9 K (see Fig. 6).

This example clearly demonstrates that a variable temperature analysis is of paramount importance when the organometallic molecule is flexible. In general a congruent decrease in the atomic displacements on decreasing the temperature reflects a true dynamic phenomenon, otherwise the atomic displacements may reflect static disorder. The low-temperature measurements also allow crystallographic data of higher accuracy to be obtained which, in turn, allow the study of the rigid body motion of the whole molecule or ions, or of molecular fragments, in terms of the T (translation), L (libration), and S (screw) tensors

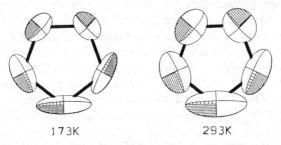

173K   293K

Fig. 6. An example of static disorder: note how the anisotropic displacement parameters of the cyclopentadienyl ligands in monoclinic ferrocene do not change appreciably on decreasing the temperature to 173 K (from [19])

[20]. These parameters provide information on the extent and direction of molecular librations and translations and on the coupling between these two motions (screw motion). The components of T, L and S can be obtained by a linear least-squares fit to the observed anisotropic displacement parameters. The dependence on temperature of these components and the presence of additional motional freedom on top of the rigid body motion can also be estimated. Furthermore intramolecular contributions to motion may be detected: the differences between the observed displacement parameters and those calculated on the basis of the rigid-body motion parameters can be used to describe the internal motions arising from "soft" vibrational and librational modes due to bond bending or other atomic displacements subjected to small restoring forces [21].

Accurate measurements of the diffraction intensities are essential to obtain reliable values for the atomic displacement parameters as errors arising from all random and systematic imperfections of the structural model will tend to concentrate in these parameters. Absorption problems, for example, are particularly important because of the presence of metal atoms in organometallic complexes and become crucial when dealing with transition-metal clusters.

2.5
Empirical Estimate of Reorientational Barriers

While in solution the molecular surroundings is made up of solvent molecules in rapid tumbling motion, molecules in the crystal are surrounded by themselves. In the case of co-crystals or of ionic crystals this applies to the repeating unit as a whole, but still with a fixed and ordered distribution of molecules or ions. Dynamic processes are therefore mainly under intermolecular control, i.e. under the control exerted by the network of intermolecular interactions that are responsible for crystal cohesion (see below).

When spectroscopic information is not available, or a complementary estimate of potential energy barriers to reorientation is sought, empirical methods widely employed in the field of organic solid state chemistry can be used [22]. The pairwise atom–atom potential energy method [23] is still one of the most useful tools to evaluate to what extent molecules or molecular fragments are held in place by the molecular or ion distribution within the crystalline edifice. The method is very well established and need not be described in the context of this contribution and the interested reader is referred elsewhere [22, 23] for details. The basic assumption is that the intermolecular interactions in a molecular crystal result from the sum of short-range repulsive and long-range attractive interactions of the kind used to describe the interaction between two isolated atoms in the gas phase. The central problem is that of the choice of potential parameters to describe adequately intermolecular interactions in the solid state. Such parameters are usually obtained by fitting observed crystal properties, such as sublimation enthalpies, by ab initio calculations of the intermolecular potential energies and/or by a statistical analysis of equilibrium atom–atom distances from data collected in databases such as the CSD [24].

One of the major limitations in the application of the pairwise atom–atom potential energy method to organometallic crystals is the lack of specific atomic parametrization for metal-containing systems. Hence, while potential parameters obtained for organic molecules can be used for light atoms, the metal atom contribution is either neglected or approximated by adopting the parameters available for the corresponding noble gases. Clearly, this type of calculation can not be expected to afford more than approximate crystal potential energy values. However, atom–atom calculations can be used as a convenient tool to investigate the spatial distribution of the molecules in the solid and to estimate the potential energy barrier to molecular or fragment reorientation (see Fig. 7) [25]. Since the barrier height is obtained as the difference between the potential energy values of the observed structure and that calculated at various reorientational steps around a predefined rotation axis (e.g. an inertial axis or a ligand-to-metal coordination axis) the problem of an exact calculation of the cohesive energy is less relevant as it can be assumed that both observed and calculated values are affected to the same extent by the approximations of the method. Reorientations can be performed either within the "static environment" approximation (thus yielding an upper limit for the barrier) or within a "cooperating" environment in which molecules of the surroundings are allowed small torsional and translational motions in order to "give way" to the reorientating molecule or fragment. The height of the potential barrier depends on the temperature at which the crystal structure is determined. These aspects of the method should always be kept in mind when comparing potential barriers with activation energies obtained from NMR spectroscopy which are obtained as mean values measured over a temperature range. In spite of the many limitations the atom–atom potential energy method has been applied with success to a number of organometallic crystals. It is worth stressing on closing this section that the atom–atom method is strictly "crystallographic" in nature as the only input required to perform barrier calculations is that of atomic coordinates, unit cell parameters and space group symmetry.

3
Organometallic Crystal Isomerization, Phase Transitions and Polymorphism

Thus far we have discussed solid state dynamic processes at a molecular level emphasizing what can (and cannot) be expected from X-ray diffraction experiments in terms of information on molecular dynamic behavior. The second part of this chapter will focus on crystal dynamics, viz. on transformations affecting the crystal edifice *as a whole*. In keeping with the general philosophy, we will concentrate on diffraction experiments and on the information that can be expected from the study of organometallic polymorphism and crystal-to-crystal transformations.

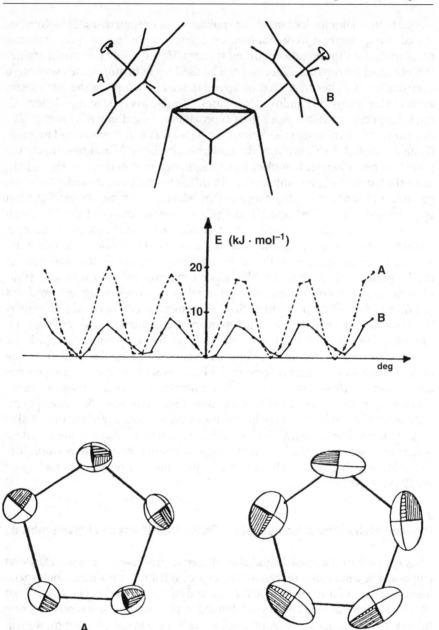

Fig. 7. Comparison of the barriers to reorientation for the motion of the ligands in crystalline cis-$(C_5H_5)_2Fe(CO)_4$: the two independent cyclopentadienyl ligands experience different crystalline environments

3.1
Crystal Polymorphism

Crystal polymorphs are alternative packing modes for the same molecule [26]. According to McCrone the number of polymorphic modifications available for any given structure is proportional to the time (and money) spent in searching for them [27]. However, while polymorphism can be a crucial problem in the pharmaceutical industry (many drugs exist in polymorphic modifications which have different efficiency of assimilation because of the difference in solubilities [28]) experiments aimed at the preparation of organometallic polymorphs have never been conceived. On the contrary, the characterization of a different crystal structure for an already known (and often already published) molecule is, most often, utterly disappointing for chemists and crystallographers having to deal with busy diffractometer queues. In the end most polymorphic modifications are not published and the knowledge of their existence lost.

The number of polymorphic structures for organometallic compounds in the Cambridge Structural Database is slightly lower than that for pure organics. The 1997 version of the CSD contains the same number of organic and metal organic molecules (including coordination complexes). However, among polymorphic forms, the pure organics occur to the extent of 63% while coordination and orgametallic complexes account for the remaining 37% [29].

Within a broader, "supramolecular" perception of the nature of a crystal [30], molecular crystal polymorphism can be seen as a form of crystal isomerism: as the different distribution of chemical bonds for molecules of identical composition gives rise to molecular isomers (e.g. *cis*- and *trans*-isomers), different distributions of intermolecular bonds give rise to isomers of the same molecular aggregate. Hence, the change in crystal structure associated with an interconversion of polymorphs, i.e. a solid–solid phase transition (between ordered phases), in which intermolecular bonds are broken and formed, can be regarded as the *crystalline* equivalent of an isomerization at the molecular level. In general, polymorph interconversion depends on whether the polymorphic modifications form an enantiotropic system (the polymorphs interconvert before melting because the solid–solid transition between polymorphs is below the solid–liquid transition) or a monotropic system (the polymorphs melt before the solid–solid transition can occur). The chemical and physical properties of the crystalline material can change dramatically with the solid state transformation.

Table 1 summarizes the different types of polymorphic and pseudo-polymorphic relationship between organometallic crystals that will be discussed here. It should be made clear, however, that this classification has only the practical purpose of description.

While polymorphs of rigid organometallic molecules are quite rare, structural flexibility plays a particularly important role in organometallic polymorphism. Structurally non-rigid organometallic molecules are likely candidates for the formation of conformational polymorphs. Many crystals of globular organometallic molecules have been shown to undergo phase transitions, and for

Table 1. Polymorphism and pseudo-polymorphism

I *Rigid molecules in different crystals*	The molecular structure is essentially unaffected by the change in crystal structure
II *Conformationally flexible molecules in different crystals*	The molecular structure is affected along soft deformational paths by the change in crystal structure
III *Structural isomers in different crystals*	Less stable structural isomers of a fluxional process are isolated thanks to the compensatory effect of external interactions in the solid
IV *Disordered pseudo-polymorphs*	Ordered and disordered crystals of the same molecules have different crystallographic symmetry arising from disorder
V *Pseudo polymorphism arising from cocrystallization*	The same molecule or ion may crystallize in different crystals with a different type/number of solvent molecules or counterions
VI *Pseudo-polymorphism of electronic isomers*	Isostructural molecules with different types and number of metal atoms

some the formation of plastic phases characterized by short-range orientational disorder and long-range order is known [31]. *Conformational polymorphism* refers to the crystallization of conformers that differ little in structure since the distribution of chemical bonds is maintained [32]. A classical example of organometallic conformational polymorphism is provided by ferrocene, for which one room-temperature disordered and two low-temperature ordered crystalline forms are known (see Fig. 8). At the crystal level they differ in the relative orientation of the molecules, so that the phase transition mechanism requires only low-energy reorientation of the rings and a limited motion of the molecules in the crystal structure [33].

3.2
Crystals Formed by Structural Isomers

An intriguing case of organometallic *pseudo*-polymorphism is that of crystals formed by structural isomers related by a low-energy interconversion pathway, e.g. cluster carbonyls with different distributions of bridging and terminal ligands [34]. The structural isomers correspond to different energetic minima along the interconversion pathway and the cohesion of the respective crystals may stabilize the less thermodynamically stable isomers. Crystals of structural isomers may (or may not) interconvert via a phase transition. In these cases, it would be highly desirable to be able to discriminate the factors controlling crystal cohesion from those controlling molecular structure, i.e. to prise apart "external" from "internal" contributions to the molecules-in-crystal global energy.

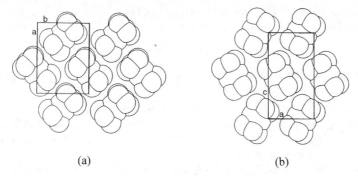

(a)   (b)

Fig. 8. Relationship between the monoclinic and orthorhombic polymorphic forms of ferrocene

The comparison of the isomer energies on a scale of relative (enthalpic) stability with the energies of cohesion of the corresponding crystals would be a reasonable approach to separate the relative contributions of either nature and to evaluate the contribution of intermolecular interactions to the stabilization of polymorphic forms. However, even the most up-to-date theoretical tools are generally not sufficient for the complexity of organometallic molecules and solids. Some insight has been obtained by combining molecular orbital analysis based on extended-Hückel theoretical calculations [35] with an analysis of the intermolecular bonding based on packing potential energy calculations [36].

3.3
Pseudo-Polymorphism of Electronic Isomers

Another type of pseudo-polymorphism is related to the intriguing possibility of changing the chemical composition without changing the distribution of chemical bonds. This is a fairly common phenomenon in organometallic chemistry as structurally similar but chemically different complexes can be obtained by replacement of one or more metal atoms with other metals in the same group (e.g. a kind of isomorphous substitution). Crystals of isoelectronic and isostructural species may crystallize in the same space group, thus being isomorphous, or in different space groups, forming *pseudo-polymorphs*.

Orthorhombic ruthenocene [37] can be seen as a pseudo-polymorph of monoclinic ferrocene. Another example is $Fe_3(CO)_{12}$ and the pair of isostructural species $Ru_3(CO)_{12}$ and $Os_3(CO)_{12}$ [38]. The former molecule possesses two bridging and ten terminal CO ligands and crystallizes in a different space group from that of the two heavier clusters, which are otherwise isostructural with twelve terminal CO's and isomorphous in their crystals.

Examples of organometallic pseudo-polymorphs in which the structures differ for the ligand distributions are the three crystal forms of $[Ru_6C(CO)_{17}]$ [39], whereas compounds of the form $[(C_5H_5)_3M_3(CO)_3]$ (M=Co, Rh and Ir) constitute a family of isoelectronic isomers [40].

3.4
Pseudo-Polymorphism Arising from Disorder

The existence of different crystal forms may also be caused by disorder, viz. the same molecules may crystallize, depending on the crystallization conditions, in different crystal forms with or without static disorder arising from different orientations of the same groupings through the crystal structure. Ferrocene dicarboxylic acid [41] affords an intriguing case of crystal pseudo-polymorphism associated with the presence (monoclinic form) or absence (triclinic form) of orientational disorder of the -COOH groups. The monoclinic form can only be obtained from hot solutions, whereas the ordered form is obtained at room temperature. Differences in entropy have been invoked to explain the phenomenon [41].

It should be stressed, however, that the presence of disorder does not per se indicate the existence of different patterns of intermolecular bonds (which is what identifies crystal isomers, viz. polymorphs). Since disorder results from the average over space (and over time in the case of dynamic disorder), the same average image of the crystal may result from superposition of crystallites with dimensions of a few nanometers (mosaic disorder) but in different orientation, or from overlap of a random distribution of unit cells containing molecules in different orientations (local disorder). In the former case the environment of each molecule is the same as in an ordered crystal.

Although organometallic crystals afford a wealth of possible combinations of molecules and crystals isomers, polymorphs of the type I are not so common. It should be kept in mind, however, that while studies concentrated on the solid state have a long-standing tradition in organic solid state chemistry, this is not so for organometallics. In our opinion the phenomena discussed above have simply been less studied. We anticipate that, as the interest in organometallic crystals grows along with that in materials chemistry, the phenomenon of crystal polymorphism in all its different aspects will attract more interest and, possibly, lead to new interesting discoveries.

3.5
Crystal Transformations and Dynamics

Many crystallographic laboratories are now being equipped with area detector diffractometers and with low-temperature devices. The collection of X-ray data at a temperature other than ambient is becoming common practice as it is being realized that more accurate data (and faster data collections) can be obtained by reducing atomic displacements. Furthermore data can now be collected on rapidly decaying crystals, extremely air-sensitive materials and even on systems which are not solid at room temperature. This represents an important improvement in crystallographic work and it is easy to predict that in the next few years not only will the number of single crystal X-ray structures characterized increase at a very rapid pace, but also that the average quality of the data, and therefore the reliability of the structural information, will also increase. This progress brings with it, however, some new problems that need to be understood.

For example, the crystalline material under investigation may undergo phase transition on a change in the temperature, i.e. belong to an enantiotropic system of the type discussed above. In some cases the phase transitional behavior of the crystal may be even more interesting than the molecular structural features per se. It is also true, however, that this could cause problems for the chemist as it may complicate, and at times slow down, the process of structural characterization (and publication!).

Often crystals are put directly onto a diffractometer at low temperature (to save the time and trouble of searching a unit cell at room temperature) and then waiting until the desired temperature is reached. This is an unwise procedure because it may mask the occurrence of a phase transition. Except in a few fortunate cases (see below) the transition may cause the crystal to break apart making indexing and cell attribution very difficult if at all possible. In most laboratories, considering the long queue of crystals to collect, this would automatically mean moving on to the next crystal. It is wise to accompany and, when possible, even to precede X-ray data collection at low temperature with an analysis of the thermal behavior of the crystalline material by means of differential scanning calorimetry (DSC). A DSC scan allows the identification of possible phase transitions and therefore to set up the diffraction experiment adequately.

Many examples of organometallic crystal transformations are available. Thiophene chromium tricarbonyl undergoes a phase transition at 185 K with the crystal "exploding" as the temperature is decreased [42]. Substituted ferrocene derivatives such as $[(\eta^5-C_5H_5)(\eta^5-C_5H_4CHO)Fe]$ and $[(\eta^5-C_5H_5)(\eta^5-C_5H_4CMeO)Fe]$, as well as salts of the type $[(\eta^5-FC_6H_5)(\eta^5-C_5H_5)Fe][A]$ $[A=AsF_6^-, PF_6^-, SbF_6^-$ or $BF_4^-]$, are all known to undergo order-disorder phase transitions [43]. In the case of the family of trinuclear cluster molecules $[Fe_3(CO)_{12}]$, $[Fe_2Ru(CO)_{12}]$ and $[Fe_2Os(CO)_{12}]$ and $[FeRu_2(CO)_{12}]$ variable temperature X-ray diffraction experiments have demonstrated the dynamic nature of the disorder associated with fully reversible order-disorder phase transitions [44]. In the case of $[Fe_2Ru(CO)_{12}]$, for instance, the disordered room temperature structure becomes fully ordered at 220 K while, on increasing the temperature to 313 K, the crystal appears to undergo a phase transition to a centrosymmetric crystal becoming isomorphous with crystalline $[Fe_3(CO)_{12}]$.

The crystalline salts $[(\eta^5-C_5H_5)_2M][PF_6]$ (M=Co, Fe) are isomorphous at room temperature [45]. The two crystals have been shown by variable temperature X-ray diffraction experiments and DSC to undergo two fully reversible phase changes (M=Fe, ca. 213 and 347 K; M=Co, ca. 252 and 314 K). Therefore the stability of the intermediate room-temperature phase is different in the two salts, i.e. ca. 62° in the case of Co and 134° in the case of Fe. These crystalline systems are remarkably robust so that the phase transitions can be followed on the single crystal diffractometer and diffraction data can be collected on the same crystal specimen. On cooling, the ordered room-temperature monoclinic crystal transforms into another ordered monoclinic crystal (with a different β-angle) with different relative orientations of the two independent cations. On heating, the crystals of the two species transform into *semi-plastic* systems containing ordered PF_6^- anions and orientationally disordered $[(\eta^5\text{-}C_5H_5)_2M]^+$ cations.

The whole phase transitional process, from the low-temperature ordered phase to the high-temperature disordered one has been interpreted as a *progressive adjustment* of the crystal edifice to the dynamic requirements of the globular organometallic cations which require more space as the temperature increases. The high-temperature phase can be described either as a *semi*-plastic cubic crystal in which the cobalticinium cations have lost long-range order and occupy the center of a "box" defined by ordered PF_6^- anions or as a monoclinic crystal (with the β-angle very close to 90°) in which the cations show a different extent of orientational disorder over two crystallographic independent sites. This latter model has the advantage of providing a path for returning to the non-degenerate ordered distribution on cooling the crystal back to room temperature. The phase transitional behavior is shown schematically in Fig. 9.

In both crystals, charge-assisted $CH^{\delta+}\text{--}F^{\delta-}$ hydrogen-bonding interactions between the C_5H_5 ligands and the PF_6^- anions appear to play an important role. CH–F bonds, though weaker than conventional OH–O hydrogen bonds, are comparable in length with charge-assisted $CH^{\delta+}\text{--}O^{\delta-}$ interactions [46].

4
Perspectives of an Organometallic Solid State and Materials Chemistry

As pointed out in the Introduction, structural organometallic chemistry is still largely dominated by the "molecular chemistry" approach, i.e. the focus of in-

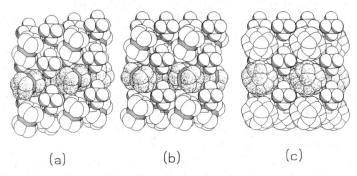

(a)   (b)   (c)

Fig. 9. Space-filling representation of the phase transitional behavior of crystalline $[(\eta^5\text{-}C_5H_5)_2Co][PF_6]$: the crystal undergoes two fully reversible transitions from an ordered low-temperature monoclinic phase (*a*) to an ordered room-temperature monoclinic phase containing cations in different relative orientation (*b*) to a high-temperature disordered phase containing ordered anions and orientationally disordered cations (*c*) (from [45])

vestigation is still on the structure of the individual reactant or product rather than on that of the solid aggregate. The reason for this is simple: alternative structural and analytical tools that may compete with X-ray diffraction in providing a rapid and detailed description of the structure of a complicated molecule or ion containing many different types of atomic species do not exist. Although it is likely that such "analytical" use of diffraction experiments will persist for some time to come, new lines of research, connected to the mainstream of synthetic and structural organometallic chemistry, are being explored.

Crystalline organometallic complexes and metal clusters represent an important bridge between molecular solids of the organic type and inorganic solids and bulk metals. Therefore, the possibility of combining the intra- and intermolecular bonding features of the ligands with the variable valence state and magnetic behavior of transition-metal atoms is an appealing aspect of this branch of chemistry. Indeed only the imagination of the experimentalists limits the way in which type, number, geometry, functionality, and electronics of ligands and metals may be combined with the variety of crystallization conditions (temperature, solvent, counterion choice, etc.) in the quest for new solid materials. The knowledge of chemical functionality at molecular level can be used to develop strategies to control supramolecular aggregation in the solid state. The approach can be theoretical or practical. In the former case crystal structures are generated computationally from known molecular structures [47] and the theoretical polymorphs compared with experimental structures. Some attempts in this directions have also been made in the organometallic field with encouraging results [48]. Alternatively (or complementarily), organometallic crystal structures can be constructed by choosing and combining adequate building blocks. Although a discussion of this topic is beyond the scope of this contribution, it is worth stressing that the utilization of crystal-directed synthetic strategies in organometallic chemistry promises to become an extremely fruitful field of research. This is what is nowadays termed "crystal engineering" [49], i.e. the intelligent utilization of non-covalent bonds to make crystals with a purpose [50]. Strong and directional hydrogen bonds, for example, can be utilized to fix organic building blocks in rigid skeletons leaving channels and cavities [51] in which organometallic molecules or ions can be accommodated or to prepare chiral supramolecular architectures in which organometallic zwitterions or strongly dipolar molecules can be oriented. The most appealing areas of application are those of optoelectronics, in particular SHG generation for optical devices [52], conductivity and superconductivity [53], charge-transfer and magnetism [54]. It is easy to predict that organometallic chemists will soon play a prominent role in the development of these areas.

Acknowledgments. We have benefited enormously from collaborations with a number of scientists worldwide and from the efforts of many students and visiting scientists at Bologna. All these exchanges have been made possible by a number of funding agencies and bilateral projects (CNR, CRUI, Vigoni, Erasmus, NATO and Ciba-Geigy exchange programs).

References

1. (a) Lehn JM (1995) Supramolecular chemistry: concepts and perspectives. VCH, Weinheim; (b) Dunitz JD (1996) In: Desiraju GR (ed) Perspectives in supramolecular chemistry. The crystal as a supramolecular entity. Wiley, Chichester
2. Kitaigorodsky AI (1973) Molecular crystal and molecules. Academic Press, New York
3. Braga D, Grepioni F, Calhorda MJ (1998) In: Braunstein P, Oro LA, Raithby PR (eds) Metal clusters in chemistry. VCH, Weinheim (in press)
4. (a) Braga D, Grepioni F (1994) Acc Chem Res 27:51; (b) Braga D, Grepioni F (1997) Comments Inorg Chem 19:185
5. Dunitz JD (1979) X-ray analysis and the structure of organic molecules. Cornell University Press, Ithaca London
6. Cotton FA, Wilkinson G (1993) Advanced inorganic chemistry, 5th edn. Wiley, New York. In this article the term 'organometallics' is used with a certain degree of freedom to include also binary carbonyl complexes and some coordination compounds which, although they do not contain a M–C bond with an 'organic' carbon, have solid state group properties which are similar to those of organometallic species
7. (a) Allen FH; Davies JE, Galloy JJ, Johnson O, Kennard O, Macrae CF, Watson DGJ (1991) Chem Inf Comp Sci 31:204; (b) Allen FH, Kennard O (1993) Chemical Design Automation News 8:31
8. Orpen AG, Brammer L, Allen FH, Kennard O, Watson DG, Taylor RJ (1989) Chem Soc Dalton Trans [Suppl] S1–S83
9. Cotton FA, Hanson BE (1980) In: De Mayo P (ed) Rearrangements in ground and excited states. Academic Press, New York, p 379
10. Braga D (1992) Chem Rev 92:369
11. We are obviously talking of single crystal X-ray diffraction techniques routinely done in thousands of chemical laboratories around the world; more sophisticated and much less accessible applications of X-ray diffraction such as those for liquids, glasses, time-resolved experiments, neutron diffraction, synchrotron radiation and applications to microcrystals, etc. are beyond the scope of this contribution
12. (a) Fyfe C (1983) A Solid state NMR for chemists. CFC Press, Guelph, Ontario, Canada; (b) Fyfe CA, Wasylishen RE (1987) In: Cheetham AK, Day P (eds) Solid state chemistry techniques. Clarendon Press, Oxford, p 190; (c) Etter MC, Hoje RC, Vojta GM (1988) Crystall. Rev 1:281
13. Giacovazzo C (ed) (1992) Fundamentals of crystallography, IUCr Series. Oxford University Press, Oxford
14. Coppens P, Hall MB (eds) (1982) Electron distributions and the chemical bond. Plenum Press, New York
15. (a) Braga D, Koetzle TF (1987) J Chem Soc Chem Commun 144; (b) Braga D, Koetzle TF (1988) Acta Crystallogr Sect B 44:151
16. Willis BTM Pryor A W (1975) Thermal vibration in crystallography. Cambridge University Press, Cambridge
17. Muetterties EL (1965) Inorg Chem 4:769
18. Benfield RE, Braga D, Johnson BFG (1988) Polyhedron 7:2549
19. (a) Takusagawa F, Koetzle TF (1979) Acta Crystallogr Sect B 35:1074; (b) Seiler P, Dunitz JD (1979) Acta Crystallogr Sect B 35:1068; (c) Seiler P, Dunitz JD (1979) Acta Crystallogr Sect B, 35:2020; (d) Seiler P, Dunitz JD (1982) Acta Crystallogr Sect B 38:1741
20. (a) Dunitz JD, Schomaker V, Trueblood KN (1988) J Phys Chem 82:856; (b) Dunitz J D, Maverick EF, Trueblood KN (1988) Angew Chem Int Ed Engl 27:880; (c) Hummel W, Raselli A, Bürgi HB (1990) Acta Crystallogr Sect B 46:683
21. (a) Braga D, Bürgi HB, Grepioni F, Raselli A (1992) Acta Crystallogr Sect B 48:428; (b) Braga D, Grepioni F, Johnson BFG, Lewis J, Housecroft CE, Martinelli M (1991) Organometallics 10:1260

22. Gavezzotti A, Simonetta M (1981) Chem Rev 82:1
23. Pertsin AJ, Kitaigorodsky AI (1987) The atom-atom potential method. Springer, Berlin Heidelberg New York
24. (a) Gavezzotti A, Filippini G (1993) Acta Crystallogr Sect B 49:868; (b) Gavezzotti A, Filippini G (1995) J Am Chem Soc 117:12299; (c) Gavezzotti A, Filippini G (1994) J Phys Chem 98:4831
25. (a) Braga D, Grepioni F, Sabatino P (1990) J Chem Soc Dalton Trans 3137; (b) Braga D, Grepioni F (1991) Organometallics 10:1254; (c) Braga D, Grepioni F (1991) Organometallics 10:2563
26. Gavezzotti A (1994) Acc Chem Res 27:309; Dunitz J, Bernstein J (1995) Acc Chem Res 28:193 and references cited therein
27. McCrone WC (1965) In: Fox D, Labes MM, Weissemberg A (eds) Polymorphism in physics and chemistry of the organic solid state, vol II. Interscience, New York, p 726
28. Byrn SR (1982) In: Solid state chemistry of drugs. Academic Press, New York, p 79
29. Braga D, Grepioni F, Desiraju GR (1998) Chem Rev 98:1375
30. Desiraju GR (ed) (1996) Perspectives in supramolecular chemistry. The crystal as a supramolecular entity, Wiley, Chichester
31. (a) Parsonage NG, Staveley LAK (1978) Disorder in crystals. Clarendon Press, Oxford; (b) Pierrot M (ed) (1990) Structure and properties of molecular crystals. Elsevier, Amsterdam
32. (a) Bernstein J, Hagler AT (1978) J Am Chem Soc 100:673; (b) Bernstein J (1987) In: Desiraju GR (eds) Organic solid state chemistry. Elsevier, Amsterdam, p 471; (c) Bernstein J (1990) In: Pierrot M (ed) Structure and properties of molecular crystals. Elsevier, Amsterdam
33. (a) Dunitz JD (1995) Acta Cryst Sect B 51:619; (b) Dunitz JD (1993) In: Kisakürek MV (ed) Organic chemistry: Its language and its state of the art. Verlag Basel, HCA, p 9; (c) Braga D, Grepioni F (1992) Organometallics 11:711
34. (a) Braga D, Grepioni F (1993) J Chem Soc Dalton Trans 1223; (b) Braga D, Byrne JJ, Calhorda MJ, Grepioni F (1995) J Chem Soc Dalton Trans 3287
35. (a) Hoffmann R (1963) J Chem Phys 39:1397; (b) Hoffmann R, Lipscomb WN (1962) J Chem Phys 36:2179; (c) Ammeter JH, Bürgi H-J, Thibeault JC, Hoffmann RJ (1978) J Am Chem Soc 100:3686; (d) Mealli C, Proserpio DM (1990) J Chem Ed 67:39
36. See for example: Braga D, Dyson PJ, Grepioni F, Johnson BFG, Calhorda M (1994) J Inorg Chem 33:3218; Braga D, Grepioni F, Tedesco E, Calhorda MJ, Lopes P (1995) J Chem Soc Dalton Trans 3297; Braga D, Grepioni F, Wadepohl H, Gebert S, Calhorda MJ, Veiros LF (1995) Organometallics 14:5350
37. (a) Seiler P, Dunitz JD (1980) Acta Crystallogr Sect B 36:2946; (b) Takusagawa F, Koetzle TF (1980) ACA Symposium, Eufaula Alabama Abstract D4, p 16
38. (a) Wei CH, Dahl LF (1969) J Am Chem Soc 91:1351; (b) Cotton FA, Troup JM (1974) J Am Chem Soc 96:4155; (c) Braga D, Grepioni F, Johnson BFG, Farrugia J (1994) J Chem Soc Dalton Trans 2911; (d) Churchill MR, Hollander FJ, Hutchinson JP (1977) Inorg Chem 16:2655; (e) Churchill MR, DeBoer BG (1977) Inorg Chem 16:878
39. (a) Sirigu A, Bianchi M, Benedetti E (1969) J Chem Soc Chem Commun 596; (b) Braga D, Grepioni F, Dyson PJ, Johnson BFG, Frediani P, Bianchi M, Piacenti F (1992) J Chem Soc Dalton Trans 2565
40. Braga D, Grepioni F, Wadepohl H, Gebert S, Calhorda MJ, Veiros LF (1995) Organometallics 14:24
41. Takusagawa F, Koetzle TF (1979) Acta Cryst Sect B 35:2888
42. (a) Chhor K, Bocquet JF, Lucazeau G, Dianoux A (1984) J Chem Phys 91:471; (b) Chhor K, Lucazeau G (1982) Spectroch Acta 38A:1163
43. (a) Daniel MF, Leadbetter AJ, Meads RE, Parker WG (1978) J Chem Soc Faraday Trans 74:456; (b) Sato K, Katada M, Sano H, Konno M (1984) Bull Chem Soc Jpn 57:2361; (c) Fitzsimmons BW, Sayer I (1991) J Chem Soc Dalton Trans 2907

44. (a) Braga D, Farrugia J, Grepioni F, Senior A (1995) J Chem Soc Chem Commun 1219; (b) Braga D, Farrugia J, Gillon AL, Grepioni F, Tedesco E (1996) Organometallics 15:4684; (c) Churchill MR, Fettinger JC (1990) Organometallics 9:446; (d) Farrugia LJ, Senior AM, Braga D, Grepioni F, Orpen AG, Crossley JG (1996) J Chem Soc Dalton Trans 631

45. (a) Braga D, Scaccianoce L, Grepioni F, Draper SM (1996) Organometallics 15:4675; (b) Martinez R, Tiripicchio A (1990) Acta Cryst Sect C 46:202; (c) Webb RJ, Lowery MD, Shiomi Y, Sorai M, Wittebort RJ, Hendrickson DN (1992) Inorg Chem 31:5211

46. Braga D, Grepioni F (1998) In: Tsoucaris G (ed) Current challenges on large supramolecular assemblies. Kluwer, Dordrecht NATO Science Series C 519:173; (c) Braga D, Cojazzi G, Grepioni F, Scully N, Draper SM (1998) Organometallics 17:296

47. (a) Gavezzotti A (1994) Acc Chem Res 27:309; (b) Karfunkel HR, Gdanitz RJ (1992) J Comput Chem 13:1171; (c) Gdanitz R (1992) J Chem Phys Lett 190:391; (d) Maginn S (1996) J Acta Cryst A52, C79; (e) Gavezzotti A (ed) Theoretical aspects and computational modeling of the molecular solid state. Wiley, Chichester; (f) Gavezzotti A (1996) In: Chitham AK, Inokuchi H, Thomas JM (eds) Current opinion in solid state and materials science p 501; (g) Gavezzotti A (1991) J Am Chem Soc 113:4622

48. (a) Schmidt GM (1971) J Pure Appl Chem 27:647; (b) Desiraju GR (1989) Crystal engineering: the design of organic solids. Elsevier, Amsterdam

49. (a) Braga D, Grepioni F, Orpen AG (1994) Organometallics 13:3544; (b) Braga D, Grepioni F, Tedesco E, Orpen AG (1995) J Chem Soc Dalton Trans 1215; (c) Braga D, Grepioni F, Sabatino P, Gavezzotti A (1992) J Chem Soc Dalton Trans 1185

50. Braga D, Grepioni F (1999) J Chem Soc Dalton Trans 1

51. (a) Braga D, Grepioni F, Byrne JJ, Wolf A (1995) J Chem Soc Chem Commun 1023; (b) Braga D, Costa AL, Grepioni F, Scaccianoce L, Tagliavini E (1996) Organometallics 15:1084; (c) Braga D, Costa AL, Grepioni F, Scaccianoce L, Tagliavini E (1997) Organometallics 16:2070; (d) Braga D, Angeloni A, Grepioni F, Tagliavini E (1997) J Chem Soc Chem Commun 1447

52. (a) Marder SR (1992) Inorg Mater 115; (b) Long NJ (1995) Angew Chem Int Ed Engl 34:21; (c) Marks TJ, Ratner MA (1995) Angew Chem Int Ed Engl 34:155; (d) Kanis DR, Ratner MA, Marks T (1994) J Chem Rev 94:195

53. (a) Williams JM, Wang HH, Emge TJ, Geiser U, Beno MA, Leung PCW, Douglas Carson K, Thorn RJ, Schultz AJ, Whangbo M (1987) Progr Inorg Chem 35:218; (b) Williams JM, Ferraro JR, Thorn RJ, Carlson KD, Geiser U, Wang H-H, Kini AM, Whangbo M-H (1992) Organic superconductors (including fullerenes): Syntheses, Structure, Properties and Theory. Prentice Hall, Englewood Cliffs, NJ

54. (a) Khan O (1993) Molecular magnetism. VCH, New York; (b) Khan O (1992) In: Bruce DW, O'Hare D (eds) Inorganic materials. Wiley, Chichester; (c) Gatteschi D (1994) Adv Mater 6:635; (d) Miller JS, Epstein AJ (1994) Angew Chem Int Ed Engl 33:385; (e) Miller JS, Epstein AJ (1995) Chem Eng News 73:30

Theoretical Treatment of Organometallic Reaction Mechanisms and Catalysis

Alain Dedieu

e-mail: dedieu@quantix.u-strasbg.fr
Laboratoire de Chimie Quantique, UMR 7551 CNRS/ULP, Université Louis Pasteur,
4 rue Blaise Pascal, 67000 Strasbourg, France

We review here the theoretical methods that are currently in use for studying organometallic reactions. Selected examples are taken from the field of homogeneous catalysis, with particular emphasis on olefin polymerization processes. These are chosen to illustrate the importance of the choice of the model; the various techniques that are used to characterize the transition state; the influence of electron correlation to account for weak interactions; the use of QM/MM methods to model the steric effects of bulky ligands. Special attention is paid to dynamic simulations that have recently been introduced in such studies, and to the inclusion of solvent effects.

Keywords: Theory, Organometallic reaction mechanisms, Catalysis

Topics in Organometallic Chemistry, Vol. 4
Volume Editors: J.M Brown and P. Hofmann
© Springer-Verlag Berlin Heidelberg 1999

1
Introduction

Over the past few years the continuous development of theoretical methods and the fast improvement of computational facilities have brought theoretical chemists to a position where they can reasonably tackle reactivity phenomena in organometallic chemistry. However, in dialogue with experimentalists, they are still faced with a dilemma: either they can use qualitative models and concepts that are easily transferable from one system to another, and thus form a basis for an understanding of reactivity; or they can rely on computations that are increasingly more accurate but which, in most instances, provide merely energy data. Such data are, of course, very useful, since they allow the computation of energy profiles of a process. However, this should be considered more as a computational experiment rather than as an integral part of a theory. In addition the theoretical chemist is more and more often asked to solve problems that combine size and complexity. This happens for instance in enantioselective catalysis, or when the reaction medium has to be taken into account. This last requirement is in fact related to another dilemma which theoreticians face: should they carry out a very accurate calculation in the gas phase and, in this case, what will be the relevance of the results for experimentalists; or should they take into account the effect of the solvent and then provide less accurate data?

The aim of the present chapter is therefore to assess the status and the performance of the theoretical methods and procedures that are currently used for studying organometallic reaction mechanisms. Recently, several review articles reassessing the theoretical studies that have been carried out in this field have been published [1–7]; these include homogeneous catalysis. Our emphasis here will be to provide the experimentalist with some guidance on how to use the results of these studies, how to understand them, and how to assess the limitations of the calculations. Thus this chapter should also complement previous reviews that have focused on the reliability of the methods in the description of molecular properties (geometry, bond energy, spectroscopic data) either in the organic or in the organometallic realms [8–14]. We will not review all the theoretical studies here, but rather limit ourselves to some representative examples taken from the field of homogeneous catalysis.

When dealing with organometallic reactivity, the theoretical chemist has to cope with many problems. The first of these is to provide a correct and accurate description of the molecular properties of the species that are involved in the reaction, in particular the reactants and the products. In several instances this may turn out to be a difficult task: when more than one electronic configuration contributes to the ground state; when there are many low-lying electronic states of various multiplicities; when the size of the ligands that are experimentally used is such that there is a large number of electrons to treat; when the metal has an unfilled shell of f-electrons (lanthanide and actinide complexes); when relativistic effects come into play. In addition, the theoretical chemist will have to cope with problems that are inherent to reactivity features: the looseness of transition states which is generally greater for transition-metal complexes than for organic systems; the coupling of the reaction path with pathways pertaining to fluxional processes; the competition of pathways that involve a different oxidation state of the metal. When dealing with the various effects of the solvent in organometallic chemistry, in addition to the polarity and/or the hydrogen bonding capacities of the solvent, its coordinating properties will also have to be considered. For ionic species the nature of the counterion may also be of importance.

To summarize, due to the versatile nature of organometallic complexes, an accurate treatment of their reactivity features very often remains a challenge. Thus a gradation of theoretical methods or approaches has been used. We shall now briefly review these methods.

2
Methods

2.1
Qualitative Molecular Orbital Theory

The qualitative molecular orbital theory was the first theory to be used extensively. Tribute should be given here to R. Hoffmann and his coworkers who pioneered the field of theoretical organometallic chemistry and organometallic reactivity, with an approach based on extended Hückel (EH) calculations [15]. From these calculations one can derive orbital interaction diagrams and orbital correlation diagrams [16–19], and use these diagrams to underline the critical features of any generic reaction. By using the fragment molecular orbital (FMO) theory [18–20] and/or the isolobal analogy concept [19–21] the results can then be generalized to similar reactions. It should be pointed out here that the derivation of orbital interaction diagrams and orbital correlation diagrams, as the FMO theory, are not restricted to EH calculations and that more elaborate calculations can also be used [22]. The EH calculations are probably best suited for this treatment, however, owing to their simplicity and also to the one particle nature of the Hückel theory: the effective Hamiltonian that is used does not contain two-particle interactions and the total energy can thus be expressed as the sum of the orbital energies.

Let us remind ourselves briefly of the main features of this theory. The first is to look at individual interactions between the orbitals of the reacting entities. The most important interactions take place between orbitals that are closest in energy and/or have a large overlap. The nature of these interactions is then examined: they are stabilizing if they involve two electrons, destabilizing if they involve four electrons. Another important feature for reactivity is the polarization of an orbital that will result from the mixing of orbitals lying above or below. This mixing can be brought about by an intramolecular perturbation (e.g. a change in electronegativity or a geometry perturbation) but, more importantly for our present purpose, by an intermolecular pertubation. This intermolecular perturbation will very often be the interaction of one or several orbitals from another molecule.

A striking example is provided by recent work reported by Alvarez and Allón [23]. These authors analyzed inter alia the approach of a nucleophile below the plane of a square planar d^8 ML_4 complex. Since this lowers the symmetry (or the pseudo symmetry) of the transition-metal complex, the empty p_z-orbital can mix into the d_z2-orbital (which is doubly occupied) (see Fig. 1). The sign of the mixing is such that it reduces the antibonding interaction between d_z2 and the σ-lone pair of the nucleophile. The d_z2-orbital is hybridized away from the incoming nucleophile. As a result, the nucleophilicity of the metal is increased and the system can more easily accommodate an electrophile that would come on top of the square plane.

Orbital correlation diagrams are also convenient tools when dealing with organometallic reactivity in the framework of qualitative MO theory. They can be either orbital or state correlation diagrams. The former are generally used for thermal reactions, while the latter are required to analyze photochemical reactions [17,24,25]. They both rely on the property that orbitals or states of different symmetry can cross during a reaction, whereas they avoid each other if they are of the same symmetry. Thus, when the symmetries of the orbitals in the reactants are matched by the symmetries of the orbitals in the product the reaction

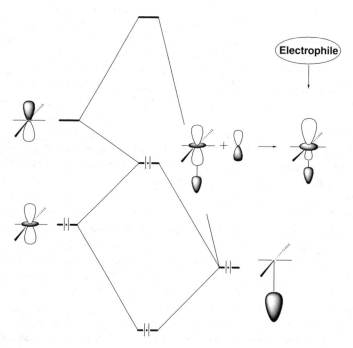

Fig. 1. Schematic orbital interaction diagram showing the second order mixing of the p_z-orbital into the d_z2-orbital, mediated by the σ-lone pair of the incoming nucleophile

is symmetry allowed. If at least one symmetry does not match, then the reaction is forbidden.

2.2
Nonempirical Methods

Although qualitative MO methods can give useful answers about reactivity problems, they cannot be used for optimizing bond lengths. Thus they cannot in most instances provide a precise determination of structures and energies of unstable intermediates and transition states. In contrast, there is now ample evidence that accurate quantum chemical methods, either (HF-based) ab initio methods or density functional theory (DFT) methods, can yield reliable results, at least for molecules of reasonable size and in the gas phase. With both types of methods a theoretical study of a reaction consists of two steps.

In step 1 the geometry of all relevant intermediates and transition states are determined [26,27]. Modern computational technologies usually allow an optimization of all degrees of freedom through a gradient technique [28–31]. Basically, all the degrees of freedom are simultaneously varied until the gradient of the energy (i.e. the first derivatives of the energy) is zero. The intermediates and transition states of the potential energy surface are therefore stationary points: the forces (which are the negative of the gradient) acting on the system at these points are zero. In order to differentiate between the local minima (intermediates, reactants, products) and the transition states the matrix of the second derivatives of the energy with respect to internal coordinates (the Hessian matrix) must be computed (either numerically or better, analytically). Its diagonalization yields information about the curvature of the surface at the stationary points: the intermediates, reactants and products are local minima (the energy is minimum in all directions). They are characterized by eigenvalues of the Hessian matrix which are all positive. Conversely, the transition state corresponds to a minimum of the energy in all directions but one. There is one and only one negative eigenvalue (which corresponds therefore to an imaginary vibrational frequency). The nature of the eigenvector with the negative eigenvalue (or the normal mode of the imaginary frequency) should be examined in order to check whether the transition state connects the reactants and the products (or the intermediates of interest in the case of a multistep reaction). In some instances it may not be immediately obvious whether the appropriate transition state has been determined and the reaction path should be followed from the transition state towards the two connected local minima, the so-called intrinsic reaction coordinate (IRC) [32]. The IRC is mathematically defined as the gradient following the (steepest descent) path in a mass weighted Cartesian coordinate system, from the transition states to the two minima. Its initial direction at the TS is given by the normal mode of the imaginary frequency. Algorithms have been derived for this task [26,33,34]. They involve a sequence of small steps, each tangent to the gradient in the mass weighted coordinate system at the starting point of that step. The whole procedure is therefore quite time consuming and it has seldom been used in transition-metal chemistry. In most cases therefore the analysis is restricted to an inspection of the normal mode of the imaginary frequency.

There are some cases in which a constrained geometry optimization has to be performed by freezing several degrees of freedom according to some educated guess. Although this is not a desirable situation, since it can lead to some strange results [27,35], this may be the only possibility when the calculation is made at such a high level of treatment that a gradient optimization technique is not available or would be too time consuming. Some care should therefore be exercised when analyzing the results.

Once the geometries have been optimized, the energy can be computed (step 2). Due to the rather large size of transition-metal complexes it is customary to perform the energy calculation and the geometry determination at different levels of computational accuracy. The geometry is obtained at a lower level in step 1 and in step 2 the energy is computed at a much higher level. It has been shown over the years that this is generally a safe procedure, although it may in some instances lead to artefacts [36].

2.2.1
Standard Ab Initio Methods

When carrying out ab initio calculations [8] an attempt is made to find an approximate solution to the time-independent Schrödinger equation $H\Psi=E\Psi$. Approximations are introduced in the Hamiltonian (H) and in the wave function (Ψ). The first of these, the Born–Oppenheimer approximation, is to leave out of the equation the nuclear motion. The nuclear coordinates are taken as parameters and only the electronic Schrödinger equation is solved. The second approximation is to express the wave function in terms of products of one-electron functions, the spin orbitals. This can be a single Slater determinant for which the variational optimization of the spin orbitals is carried out in a self-consistent field (SCF) manner through the canonical Hartree–Fock (HF) equations. In practice the spin orbitals are expanded in a basis of functions, generally centered on the atoms, the so-called linear combination of atomic orbitals to molecular orbitals (LCAO-MO) procedure. The basis functions used in the most popular softwares are Gaussian-type atomic functions (i.e. they have a radial part of the form $\exp(-\alpha r^2)$). The larger the basis, the more accurate the determination of the energy will be. Moreover it may be crucial in some instances to include the so-called polarization functions. These are functions of one higher angular momentum than those in the valence orbital of the atom (e.g. p-functions on H atoms, d-functions on C, N, O, etc.). Their primary purpose is to increase the angular flexibility in the variational process, by allowing the expansion of the wave function into regions that would not be well described otherwise.

The single determinant (or single configuration) picture and the HF approximation do not account for electron correlation [8,37]. This electron correlation can be of dynamic nature, i.e. related to the dynamic motions that the electrons undergo in order to "avoid" one another at each instant in time or it can be of non-dynamic nature, e.g. related to the fact that the ground state of the system involves more than one electronic configuration, to the need of proper dissociation of covalent bonds, to the existence of configuration crossings in a reaction. To handle the electron correlation problem, a multiconfigurational description of the wave function Ψ must be resorted to. The wave function is expanded over a basis of determinants or configurations: $\Psi = \Sigma\, C_k \Phi_k$ and the Schrödinger equation is solved either variationally or pertubatively.

There are two ways to perform the variational treatment [8,37]: (i) the Φ_ks can be kept unchanged and only the C_k optimized. This corresponds to the configuration interaction (CI) methods [38]. The Φ_ks are constructed from a reference Φ_0 wave function by exciting from the doubly occupied orbitals to the empty orbitals, one, two, three, etc... electrons. We can then speak of CIS (CI with single excitations), CISD (CI with single and double excitations), CISDT (CI with single, double and triple excitations) calculations, and so on; (ii) the C_ks and the Φ_ks can also be optimized simultaneously. This corresponds to the multiconfigurational self-consistent field (MCSCF) methods; the CASSCF method [39–41] belongs to this category. To achieve convergence in the variation procedure with the MCSCF methods is rather difficult. The number of Φ_ks is therefore generally kept to a minimum, chosen so as to describe the essential correlation effects (mostly those of non-dynamic nature).

The Schrödinger equation can also be solved pertubatively [8,37]. One of the most popular perturbation methods is the Möller–Plesset (MP) method [42–44]. In this method the Hamiltonian is chosen as the sum of the Fock operator (obtained in the SCF procedure) for each of the electrons. The perturbation can be carried out to various orders, the perturbation equations determining which Φ_k to be included through a given order. In this way the MP2, MP3, MP4, etc. treatments can be found. Current computational facilities do not permit investigations beyond the 4th order for transition-metal complexes.

In order to take into account both non-dynamic and dynamic electron correlation effects, a multireference configuration interaction calculation (MR-CI) can be performed in which the reference wave function Φ_0 is no longer a single determinant but a linear combination of determinants (or of configurations). This reference wave function is usually the outcome of a MC-SCF calculation designed to take care of the non-dynamic correlation effects. The remaining dynamic correlation effects are taken into account by a CI treatment in which the excitations are produced from this multideterminantal or multiconfigurational reference wave function. The applicability of the method is hampered by the large number of excited configurations that are generated. A more convenient procedure is to use a low order perturbation treatment for the dynamic correlation effects. A popular variant is the CASPT2 method [45] where the reference wave function is a CASSCF wave function and the perturbation treatment is carried out up to the second order.

Rather than expressing the wave function as a sum of configurations or determinants, the coupled cluster (CC) method [46] expresses it as $\Psi = \exp(T)\Phi$, where Φ is usually the SCF determinant and the operator T acts on Φ to generate single, double, triple, etc., excitations. This gives rise to the acronyms CCSD, CC-SDT, and so on. The CCSD(T) method refers to a method in which the (T) accounts for a perturbational estimate of the triple excitations. For transition-metal complexes, the CCSD(T) level is currently the limit of what can be calculated computationally and is usually considered to be the state of the art. The quadratic configuration interaction (QCI) method [47] is an approximation [37] to the CCSD method and usually gives similar results. It is worth noting, however, that in these methods, as in the Möller Plesset method, the ground state wave function should be dominated by a single configuration.

One approximation that is almost systematically made in the Hamiltonian, when dealing with transition-metal complexes, is to treat explicitly only the valence electrons of the metal, and to describe the action of the inner shell electrons (that can be viewed as chemically inert) on the valence electrons by an effective core potential (ECP) [13,48]. In addition, to easing the computational problem by reducing the number of electrons to be treated, this methodology allows a straightforward introduction of most of the relativistic effects (in particular the Darwin term and the mass-velocity term) [14,49,50]. Thus relativistic effective core potential (RECP), whose parameters are derived from atomic calculations that include these terms, can be defined. Spin orbit effects are almost always ignored. This is a reasonable assumption, especially for saturated systems.

2.2.2
Density Functional Theory Methods

An approach that is becoming more and more popular is the density functional theory (DFT) [2,3,51–54]. In this method the electronic Schrödinger equation is not solved. Instead, the energy is expressed as a functional of the electron density of the system. This method takes care of the electron correlation. Many variants exist that correspond to different exchange and correlation functionals and to different combinations of these functionals. For a full analysis of DFT methods, we refer the reader to the chapter by N. Rösch. For the sake of the foregoing discussion, a brief account of local and non-local approximations for these functionals [54] will suffice. In the local density approximation (LDA) the exchange correlation energy is assumed to be a function, and not a functional, of the electron density. It is obtained by fitting procedures from the exchange-correlation energy of the uniform electron gas. The LDA approximation generally gives good results for the determination of the structural features of systems. However, it usually leads to binding energies that are too large. This can be improved by adding non-local (NL) corrections to both the exchange and correlation functionals, through terms that involve the gradient of the density. Another approach that has been quite successful is to deviate somewhat from pure DFT methods and to include, via appropriate fitting, some Hartree–Fock exchange into the DFT exchange functional. The so-called B3LYP method [55–57] belongs to this class of methods, and seems to give reasonably accurate results, especially for geometry optimization, at a computational cost somewhere between HF and MP2 calculations.

2.3
QM/MM Methods

The above methods, either standard ab initio or DFT, are restricted to systems of relatively small size, consisting typically of one or two transition metals and of ligands with a total number of lighter atoms not exceeding 30 to 40. An improvement towards the description of more realistic organometallic systems (e.g. with PPh_3 or $P(t\text{-}Bu)_3$ ligands instead of PH_3 or PMe_3) is to combine quantum mechanics and molecular mechanics calculations. In this approach, which is the subject of the chapter by Maseras, the active site of the system is treated quantum mechnically (QM) and the remainder of the system with a molecular mechanics (MM) force field. The remainder can therefore be very large. The corresponding calculations should account relatively well for non-bonded interactions (i.e. steric effects) especially if they do not involve highly polarized groups (unless the QM/MM Hamiltonian takes care of the electrostatic coupling between the QM and the MM atoms).

2.4
QM/MD Methods

The above methods are of "static" nature. Information about the fate of a reaction is given indirectly through transition state theory. The thermodynamic data (entropies and free energies) can be derived from a thermal analysis (using statistical mechanics) of the frequency calculation for the stationary states (reactants, products, intermediates, transition states) along the reaction path. However, it would be desirable to elucidate the dynamics of the system through a molecular dynamic (MD) simulation. The Car-Parrinello (CP) method [58,59] is a MD method in which both the electronic degrees of freedom and the nuclear motions are handled simultaneously during the MD simulations. The quantum description of the electronic wave function is based on DFT in the LDA approximation. Fictitious masses are assigned to the electrons and Newton equations of motion that take care of both the motion of electrons and nuclei are solved. Choosing fictitious masses for the electrons that are as small as possible and a time step in the integration scheme that is sufficiently short allows the electrons to follow the nuclear motion adiabatically. As will be shown later, successful simulations of organometallic reactions have been performed, allowing the time evolution of the geometrical parameters to be followed, e.g. bond-breaking and bond-making processes, crucial angle variations.

From a practical point of view, however, total simulation times of pure CP dynamics cannot presently exceed 1 to 2 ps (1000 to 2000 fs). This limits the study to chemical reactions with no or a very low activation barrier. As noted by Margl, Ziegler and Blöchl [60], processes with barriers greater than about 5 kcal mol^{-1}

cannot be handled in this way, since most of the events of interest do not take place or take place very rarely. In such cases the simulation accounts only for fast movements that are not of interest for the reaction. This difficulty has been circumvented by the use of a technique based on constrained CP dynamics [60]. In this technique fictitious dynamics along a chosen reaction coordinate are introduced to explore the vicinity of the transition state. The reaction coordinate which is kept constrained during the dynamics needs to be as close as possible to the intrinsic reaction coordinate (IRC). The free energy difference ΔF between two arbitrary points ($\lambda=0$ and $\lambda=1$) along the reaction path can be determined by thermodynamic integration [61] as:

$$\Delta F = \int_0^1 < \delta E / \delta \lambda >_{\lambda, T} d\lambda$$

where λ is a parameter that varies linearly between 0 and 1. Using a very slow variation of λ (slow growth technique) allows the determination of only one single value of the force $\delta E/\delta \lambda$ on the reaction coordinate for each λ. The integral $\int < \delta E/\delta \lambda >_{\lambda,T} d\lambda$ can be used to identify stationary points along the reaction coordinate: it is minimum for intermediates and maximum for transition states.

One step further in the simulation of reactions is to combine this CP methodology with the QM/MM methodology, as reported very recently by Ziegler et al. [62]. In this implementation a reaction coordinate is again constrained during the dynamics. The electronic wave function is now based on a QM(DFT)/MM description.

3
Selected Examples

All these tools and methodologies have been applied with some success to a variety of catalytic processes or at least to the analysis of crucial steps in catalytic processes. The following is a review of some of them.

3.1
Olefin Polymerization Using Metallocene-Based Transition-Metal Catalysts

The olefin polymerization homogeneously catalyzed by d^0 metallocene complexes, the analog of the heterogeneous Ziegler–Natta process, has recently experienced renewed interest [63,64]. The propagating step involves, as originally proposed by Cossee and Arlman [65,66], the insertion of the olefin into the Ti–C bond between the metal and the growing alkyl chain, see Scheme 1. The model was refined some years later by Brookhart and Green [67], who proposed that agostic interaction might assist the olefin insertion. That this insertion is a symmetry allowed process has been shown by Lauher and Hoffmann [68] and by Jolly and Marynick [69]. An orbital correlation diagram for the model reaction $[Cp_2Ti(H)(C_2H_4)]^+ \rightarrow [Cp_2Ti(C_2H_5)]^+$ reproduced according to the work of Lauher and Hoffmann [68] is shown in Fig. 2 (the fragment molecular orbitals have been drawn using the results of related work on the reactions

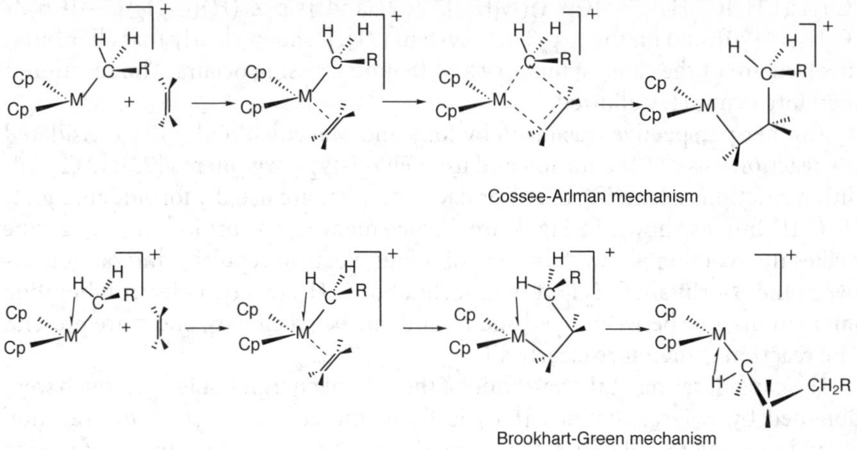

Cossee-Arlman mechanism

Brookhart-Green mechanism

Scheme 1

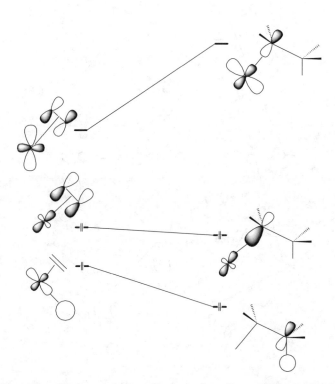

Fig. 2. Schematic orbital correlation diagram for the model reaction $[Cp_2Ti(H)(C_2H_4)]^+ \rightarrow [Cp_2Ti(C_2H_5)]^+$

$[Cp_2Ti(CH_3)(C_2H_4)]^+ \rightarrow [Cp_2Ti(C_3H_7)]^+$ [69] and $[Cp_2Zr(H)(C_2H_2)]^+ \rightarrow [Cp_2Zr(C_2H_3)]^+$ [70], and on the $[Cp_2TiR]^+$ system [71]. It shows clearly that all orbitals involved are of the same symmetry and that no crossing occurs. The reaction is therefore symmetry allowed.

Another perspective was taken by Jolly and Marynick [69]. They considered the reaction as a [2+2] addition and used Fukui-type arguments [72]: [2+2]-addition reactions occurring in a suprafacial manner are usually forbidden, e.g. H–H+C_2H_4 but, as shown in Fig. 3, the involvement of a d-orbital of $d_{x^2-y^2}$ type makes the reaction allowed: instead of a four electron repulsive interaction between the σ-orbital of H_2 and the π-orbital of C_2H_4, two two-electron bonding interactions, one between s_H+d and $π^*$ and one between $d-s_H$ and π are present. The reaction is therefore clearly allowed.

The consideration of the variation of the orbital energies of Fig. 2 (which were obtained by EH calculations [68]) leads to the conclusion that the reaction should proceed rapidly for a d^0 system: there is an energy gain in transforming the M–H and M–(C_2H_4) bonds into C–H and M–(C_2H_5) bonds. Simple thermodynamic considerations also lead to the same result. On the other hand, for a d^2

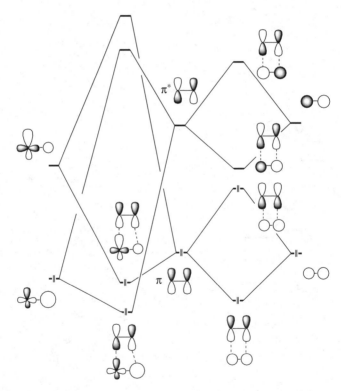

Fig. 3. Schematic orbital interaction diagram for the [2+2] suprafacial addition of the C=C double bond of C_2H_4 to the H–H bond (*right*) and to the M–H bond (*left*), respectively

system such as the neutral $[Cp_2Nb (H)(C_2H_4)]$ system, a more difficult process should be obtained since the highest orbital in Fig. 2 is occupied and destabilized during the insertion. The ab initio calculations that were performed later did not alter these conclusions.

We have already mentioned that the original mechanism of Cossee [65–66] was later modified by Brookhart and Green [67], who proposed that the insertion might be assisted by an agostic interaction of one of the α-H atoms of the growing alkyl chain. Ab initio studies to be described later in this account support this hypothesis. EH studies have also been carried out to investigate this point [73–75]. Despite the intrinsic inability of EH theory to reproduce bond lengthening and bond shortening in a quantitative way, the study of Brintzinger et al. indicated the preference for a biphasic reaction pathway made of an alkyl tilt followed by the olefin shift (Scheme 2) over a concerted pathway in which the alkyl tilt and the olefin shift would occur concomitantly [73]. The orbital analysis traced the relatively low destabilization associated with the alkyl tilt to the interaction of the C–H bonding orbital with the $5s$ and $4d_{x^2-y^2}$ orbital, see **1** and **2**. Thus, although no agostic interaction is found in the ground state reactant structure, it becomes important in the transition state and precludes the development of an important energy barrier.

Despite their usefulness in assessing most of the broad features of the homogeneous Ziegler–Natta process, the qualitative MO studies could not provide answers unambiguously and/or quantitatively to many questions:

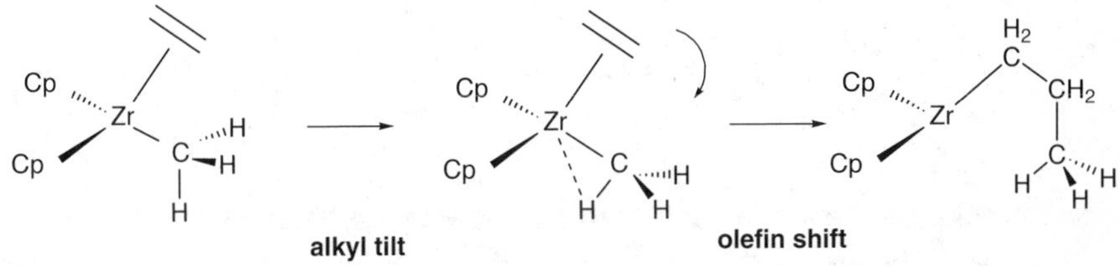

Scheme 2

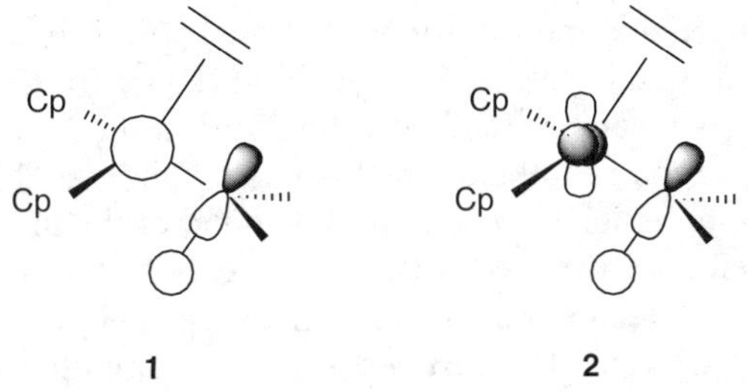

Structures 1 and 2

(i) What is the structure of the growing chain of the resting state between the successive insertions? Is it stabilized by α-, β- or γ-agostic interactions?

(ii) What is the origin of the chain propagation barrier? Is the barrier due to some rearrangement in the resting state, or to the insertion of the olefin into the metal–carbon bond?

(iii) How do the chain termination steps and chain branching steps compete with the chain propagation state?

(iv) What is the rationale for the stereoregulation in the α-olefin polymerization?

(v) What is the influence of the metal and of the first coordination sphere on the process?

To answer these questions there have been in the last few years a multitude of ab initio theoretical studies that have addressed these questions [69,76–90,96]. The geometries were optimized either at the HF, MP2 or DFT level, the energetics then being evaluated either at the MP2, DFT or at even higher levels of accuracy. More recently, molecular dynamic simulations have been carried out independently [91,92] or combined with some of the ab initio "static" calculations [93,95]. We will now review these studies highlighting in particular some of the problems that have been encountered as examples of difficulties that may arise when dealing with ab initio calculations in organometallic reactivity.

Due to computational limitations, the first ab initio studies considered for the insertion step was reaction (1) in which the Cl ligand stands as a model for the Cp ligand.

$$[Cl_2TiCH_3]^+ \; + \; C_2H_4 \longrightarrow [Cl_2TiC_3H_7]^+$$

(1)

The geometries were optimized either at the HF level (with a rather small basis set, lacking in particular polarization functions) [76,77] or at some semi-empirical level, assuming Cs symmetry throughout the entire reaction path [69]. The energetics were in some instances recomputed at the MP2 level. These calculations confirmed the results obtained at the qualitative MO theory level: a strong α-agostic interaction between the C–H bond of the methyl and the titanium atom was found in the transition state but not in the π-ethylene $[Cl_2Ti(C_2H_4)(CH_3)]^+$ precursor complex [77]. A very weak α-agostic interaction was also found in $[CH_3TiCl_2]^+$. The energy barrier computed for the insertion was modest, especially when electron correlation was taken into account:

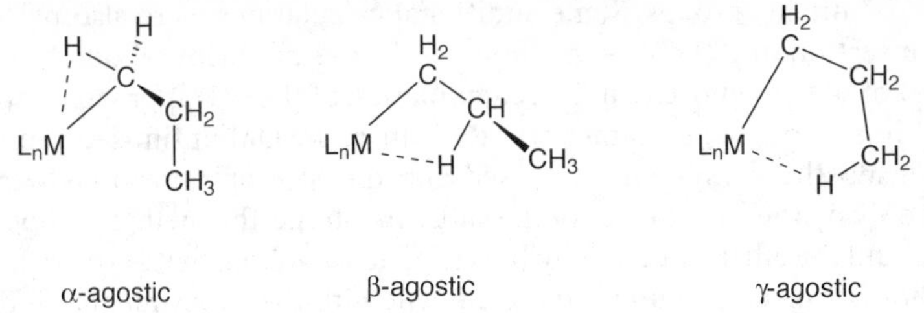

α-agostic   β-agostic   γ-agostic

Scheme 3

4.3 kcal mol^{-1} at the MP2 level. This value is in relatively good agreement with the experimental estimate for the activation energy, 6–8 kcal mol^{-1} [97–100].

As far as the product of the reaction is concerned, there are three possible types of agostic structure for the propyl ligand, see Scheme 3. The most stable structure that was found was a γ-agostic structure, with both C$^\beta$–C$^\gamma$ and C$^\gamma$–H agostic interactions with the metal. The β-agostic structure was 4.7 kcal mol^{-1} higher in energy. Interestingly, too, no α-agostic structure of the product is mentioned in a paper by Morokuma et al. [77]. The same authors extended their study to the insertion of C_2H_4 into the Zr–C bond of the silylene-bridged system $[(SiH_2)Cp_2ZrCH_3]^+$ [78] [Eq. (2)].

$$[(SiH_2)Cp_2ZrCH_3]^+ + C_2H_4 \longrightarrow [(SiH_2)Cp_2ZrC_3H_7]^+$$

(2)

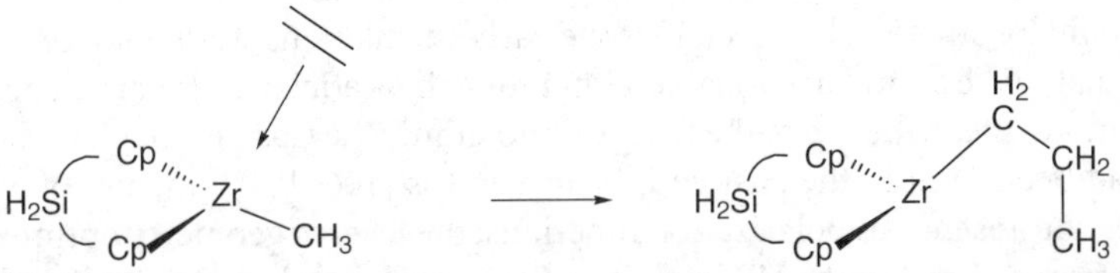

The optimization of the structure was again limited to the HF level and similar conclusions were reached. As in the model system $[Cl_2Ti(C_2H_4)CH_3]^+$ the α-agostic interaction was found to develop only in the transition state. The moderate energy barrier, 6.0 kcal mol^{-1} at the MP2 level, is also in good agreement with the experimental estimate for the propagation step [97–100]. In contrast to the previous bis-chloro system, the β-C-H agostic structure of the product is found to be more stable (by 2.5 kcal mol^{-1} at the MP2 level) than the γ-C-H agostic structure. This is already some indication that the bis-chloro system might not be a good model.

This study is also interesting because it was one of the first attempts to combine molecular mechanics (MM) calculations with a quantum mechanical (QM) treatment for investigating the regioselectivity and the stereoselectivity of the insertion of propylene into the Zr–methyl bond of the $[(SiH_2)(CpMe_n)_2 ZrC_3H_7]^+$ system, in which two or four hydrogen atoms of the two Cp rings are replaced by methyl groups. Some additional calculations were also performed for the insertion into the Zr–ethyl bond and Zr–2-methylbutyl bond to model the effect of the growing chain. The combination of the two approaches was carried out in a stepwise and rather crude manner: the QM optimized geometries of the unsubstituted system were used and the MM optimizations were performed only on the coordinates of the substituents, i.e. the methyl groups on the Cp rings and the ethyl and 2-methylbutyl ligands. Looking first at the regioselectivity, Morokuma et al. found – in agreement with the experimental findings – that the primary insertion is favored over the secondary one. The methyl substitution on the Cp rings makes the discrimination between the two pathways larger. It is worth mentioning here that these effects were found to be operative for the transition state and *not* for the π-complex precursor. This underlines the power of a QM/MM approach for such problems. The previous MM calculations were restricted to the search of equilibrium geometries [101–106]. Assessing reactivity features only on the basis of the geometries of intermediates (and not transition states) might be quite dangerous. Next Morokuma et al. analyzed (with similar QM/MM calculations) the stereoselectivity in the propylene polymerization and found an *indirect* steric control where the substituents on the Cp rings determine the conformation of the polymer chain and this conformation in turn determines the stereochemistry of the olefin insertion at the transition state [78]. Similar conclusions were reached in a study they carried out later along the same lines, but for a much larger sample of silylene-bridged group 4 metallocenes [90]. This included the C_2-symmetrical $H_2Si(indenyl)_2MMe^+$ and $H_2Si(tetrahydroindenyl)MMe^+$ (M=Ti, Zr, Hf) and the asymmetrical $H_2Si(3$-*tert*-butylCp)(fluorenyl)ZrMe$^+$ systems. In addition, various substitutions on the indenyl rings of the bis-indenyl system were examined [90].

It should be stressed, however, that the early calculations were not free of limitations that make them questionable, either for a theoretician or for an experimentalist. Among these, the use of the Cl ligands to mimic the Cps (although such modeling had been used in the past with some success [106–109]); the use of a small basis set; the absence of polarization functions; the level of geometry optimization which was restricted to the HF level. It can also not be excluded that some fortuitous cancellation of errors led to similar conclusions for reactions (1) and (2).

This seems to be indeed the case. Ahlrichs and coworkers have carried out a rather thorough study [80] of the insertion of C_2H_4 into the Ti–CH_3 bond of $[Cl_2TiCH_3]^+$ and of $[Cp_2TiCH_3]^+$, respectively, i.e. reactions (1) and (3).

$$[Cp_2TiCH_3]^+ \ + \ C_2H_4 \ \longrightarrow \ [Cp_2TiC_3H_7]^+$$

(3)

The geometries were optimized at the HF, MP2 and DFT-LDA levels, using basis sets with polarization functions on the chlorine atoms and on the carbon atoms of CH_3 and C_2H_4 only. No α-agostic interaction, either for $[Cl_2TiCH_3]^+$ or $[Cp_2TiCH_3]^+$, was found at the HF level. On the other hand, a strong α-agostic interaction was found for $[Cp_2TiCH_3]^+$, either at the MP2 or DFT-LDA level, but not for $[Cl_2TiCH_3]^+$. Thus the $[Cl_2TiCH_3]^+$ system appears to be a reasonable model of $[Cp_2TiCH_3]^+$ at the HF level only. *This result underlines the fact that weak interactions such as agostic interactions are usually not well accounted for by a HF calculation, but need to be treated at a correlated level* [110].

Whether or not strong agostic interactions are present in $[Cp_2TiCH_3]^+$ and in the growing chain polymer complex is important since it will govern its subsequent reactivity. The presence of an α-agostic interaction was put forward by Ahlrichs and coworkers to rationalize the fact that they did not find, either at the MP2 or DFT-LDA levels, the formation of a π-ethylene complex: instead, during the geometry optimization process, the $[Cp_2TiCH_3]^+ + C_2H_4$ system evolved directly into the $[Cp_2Ti\,C_3H_7]^+$ product, *without* any insertion barrier. Similar results were obtained subsequently by Ziegler et al. in their DFT study of reaction (2). The π-ethylene complex was found to lie in a very shallow minimum, the energy barrier for the insertion amounting to only 0.1 kcal mol^{-1} [82,84]. In these calculations the geometries were optimized in the local approximation, but with a polarized basis set on CH_3 and C_2H_4. The energy profile, on the other hand, was determined with non-local corrections.

The very low energy barrier computed by Ahlrichs et al. [80] and by Ziegler et al. [82,84] for the insertion step in reactions (2) and (3) contradicts somewhat the results of Morokuma et al. [77,78]. Since the local approximation in DFT calculations is known to overestimate correlation energy, it might have led to some bias in the geometry optimization. Thus Morokuma and coworkers reinvestigated more closely reaction (2) [87]. They optimized the geometries at the SCF, MP2 and DFT-NL levels (here the B3LYP level), but with an unpolarized basis set. The structures of the π-ethylene complex and of the transition state that they obtained were much closer to each other than to the DFT-LDA structure of Ziegler. In particular the α-agostic interaction was found, as in their previous calculations, only in the transition state and not in the π-ethylene complex. The energy barrier computed at the DFT-NL level, 1.7 kcal mol^{-1}, was found to be much smaller than the value computed either at the MP4 or QCISD levels (9.2 and 9.4 kcal mol^{-1}, respectively). All these results led Morokuma to question the DFT results.

It should be pointed out, however, that the determination of agostic interaction relies not only on the method, but also on the quality of the basis set. We have mentioned that the calculations of Ziegler on $[(SiH_2)Cp_2ZrCH_3]^+$ were carried out with a polarized basis set on the carbon and hydrogen atoms. In contrast, Morokuma's calculations did not include polarization functions on the hydrogen atoms. Calculations performed by Axe and Coffin [111] on the $[Cl_2TiCH_3]^+$ system at the HF, MP2 and DFT levels (DFT-LDA and DFT-NL) with a basis set of good quality for the ligands, including in particular polarization functions on the hydrogen atoms, did show the presence of α-agostic interactions. *Thus it is clear that the inclusion of electron correlation and the addition of polarization functions are both important for unambiguously assessing the presence of C–H agostic interactions.*

The above discussion may have appeared quite lengthy, if not useless, for an experimentalist. Yet, the presence of agostic interactions is a key feature of the mechanism of the homogeneous Ziegler–Natta process. Moreover, our purpose throughout this discussion has been to stress for the non-specialist that, even for semiquantitative results, relatively high level calculations are necessary in organometallic reactivity. A good knowledge of the performance of the theoretical model is also required. Such a knowledge obviously requires some careful cross checking.

The results of the DFT calculations for reaction (2) have been corroborated by the results of the first Car–Parrinello type simulation, by Meier et al. [91], of the full dynamics of an organometallic reaction. The molecular dynamics simulation started from the structure of $[(SiH_2)Cp_2ZrCH_3]^+$ without the α-agostic interaction. Two figures displaying geometrical variables as a function of simulation time are worth analysis in this review, since they are typical of the kind of information that can be obtained from such simulations. From Fig. 4, that shows the time evolution of the methyl/ethyl distance as a function of time, it is clear that the ethylene inserts into the Zr–methyl bond over a time span ranging from approximately 70 to 170 fs. This extremely fast process (at the time scale of the slow molecular vibrations in the complex) is very indicative of the very low, if any, reaction barrier.

Also noteworthy is the development of an α-agostic interaction during and not at the onset of the reaction, see Fig. 5. Analogous results have been obtained

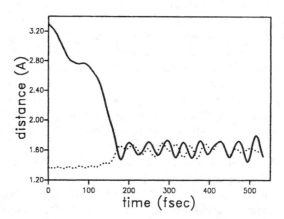

Fig. 4. Time evolution of the distance between the methyl carbon atom and the nearest ethylene carbon atom (*solid curve*), and of the ethylene internal C–C bond length (*broken curve*). (Figure reproduced with permission from [91])

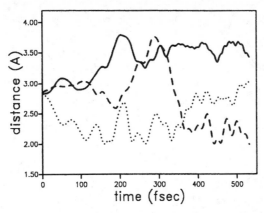

Fig. 5. Time evolution of the distance between the Zr atom and each of the three hydrogen atoms belonging to the methyl group originally bonded to Zr. The time evolution of one of the hydrogen atoms (*dotted curve*) shows the development of α-H agostic interaction. Later in the simulation (after about 450 fs) one of the other hydrogen (*broken curve*) exhibits γ-H agostic interaction. (Figure reproduced with permission from [91])

more recently in a simulation of reaction (3) [92] (although in this case the time to insertion was somewhat larger, 500 fs).

What about the fate of the direct product of the insertion? We have already mentioned that this direct product corresponds to a γ-agostic structure. However, both the ab initio and the DFT studies have shown that a β-agostic structure is more stable by a few kcal mol^{-1} [78,80,84,87] (except for the very crude model). This is in agreement with X-ray crystal structures of $Cp^*_2Sc(C_2H_5)$ [112,113]. The interconversion of the γ-agostic product into the more stable β-agostic structure has been analyzed. Morokuma and coworkers determined the transition state for this interconversion in the products of reaction (2), $[(SiH_2)Cp_2 M(C_3H_7)]^+$ (M=Ti, Zr, Hf) [87]. It corresponds to a rotation around the C^β–C^γ bond. In the titanium system there are no agostic interactions left in the transition state, whereas some are retained in the Zr and Hf systems. The loss of agostic interaction in the Ti system was traced by Morokuma to the small ionic radius of Ti which, in turn, leads to the compactness of the system and to some steric hindrance around Ti. The isomerization barrier amounts to 5.5, 4.4 and 8.6 kcal mol^{-1} (QCISD level) for Ti, Zr and Hf, respectively.

As far as the chain termination is concerned, the β-H elimination (see Scheme 4) is, according to the studies of Morokuma [87] and of Ziegler [84], quite endothermic. Values ranging between 42 and 52 kcal mol^{-1} have been computed. The barrier is also quite substantial. Thus this reaction, which would give a vinyl-terminated polymer and a metal hydride, cannot be a chain termination of the polymerization process. In fact it seems from the calculations of Morokuma and from those of Eisenstein et al. [114] that the β-Me elimination would be easier. The MO analysis of the β-Me elimination made by Eisenstein et al. [114]

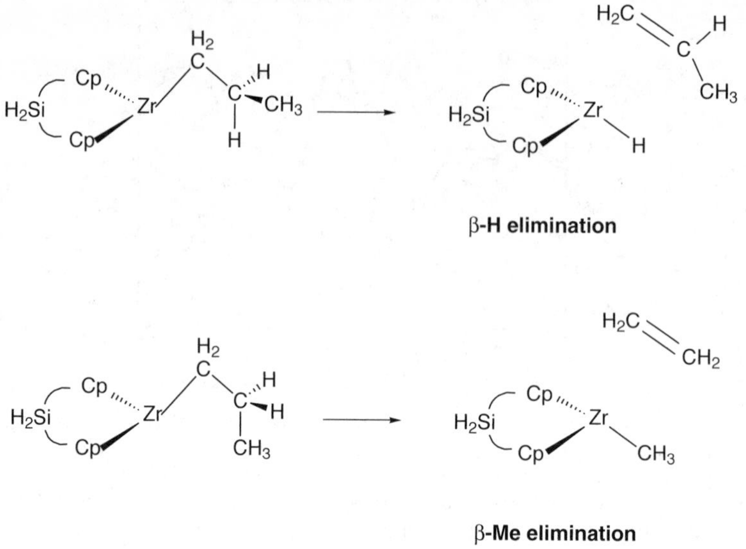

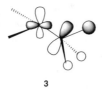

β-H elimination

β-Me elimination

Scheme 4

3

Structure 3

traces this feature to a stabilizing interaction, in the methyl product, between the doubly occupied π-orbital of Me and a d_π empty orbital on the metal, see 3.

As noticed by Ziegler [83,86], the fact that the β-agostic structure is the most stable structure of the product of the insertion makes it a good candidate for the resting state of the propagation chain. If this is true, then the use of CH_3 as a model for the growing chain in the insertion step (as for instance in reactions (1), (2) or (3)) is not appropriate, since the methyl cannot reproduce the β-agostic situation. Moreover, very low energy barriers (less than 1 kcal mol^{-1}) were computed for the insertion of C_2H_4 into the Zr–CH_3 bond (vide supra), whereas experimental investigations yielded propagation barriers of the order of 6–8 kcal mol^{-1}. Thus, the barrier for the propagation step might well be due to some rearrangement of the alkyl chain, and not to the insertion of C_2H_4 in the Zr–alkyl chain. Such a rearrangement can obviously *not* be studied with CH_3 as a model for the alkyl chain. Again we see here how the choice of the model is crucial and should be carried out with care if a relevant theoretical analysis is required.

In their most recent studies, Ziegler and coworkers therefore chose the C_2H_5 group as a more realistic model of the growing chain polymer [83,85,86,89,93–95]. They assumed that the resting state is the β-agostic conformer and considered first model reaction (4) [83], as a first step towards more elaborate studies (vide infra).

$$\left[Cl_2Ti(C_2H_5)\right]^+ \;+\; C_2H_4 \longrightarrow \left[Cl_2Ti(C_4H_9)\right]^+ \tag{4}$$

Two pathways for the approach of the olefin were considered, see Scheme 5, either the frontside attack (FS) or the backside attack (BS). Contrary to the backside attack, which can only lead to the insertion, the frontside attack can result either in the insertion or in the hydrogen exchange, see Scheme 5. For the frontside

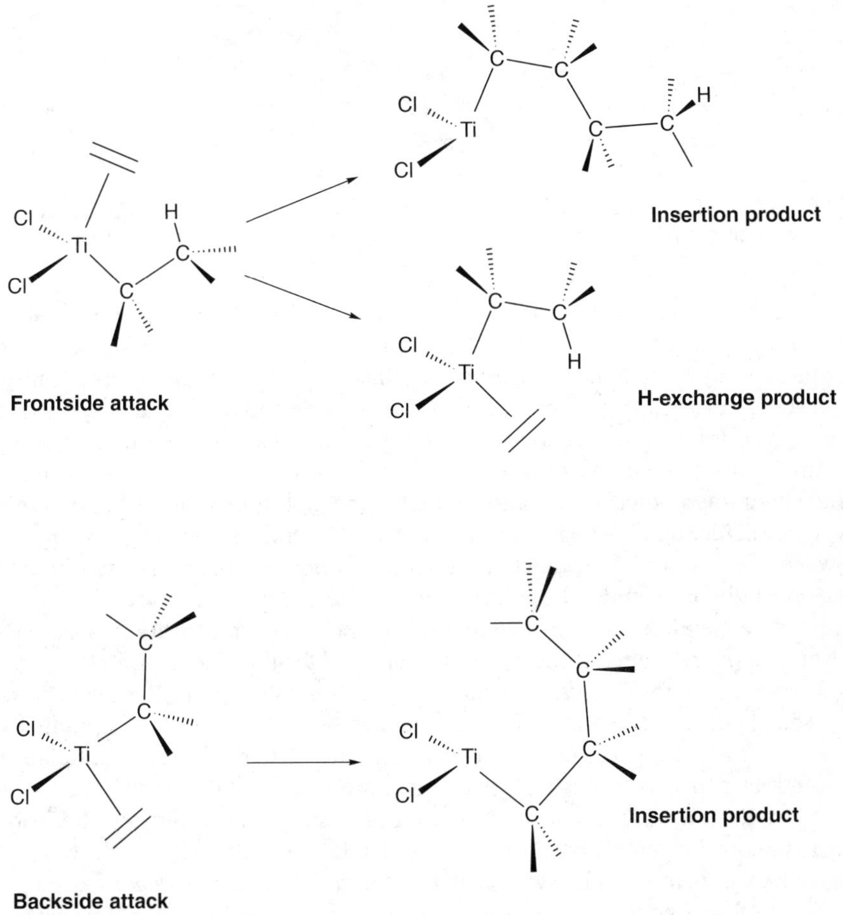

Scheme 5

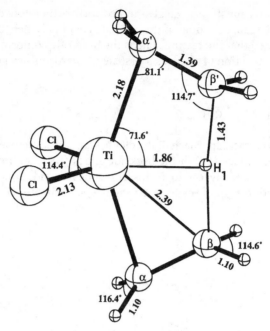

Fig. 6. Optimized (DFT-LDA level) structure of the transition state for the hydrogen exchange process in $[Cl_2Ti(C_2H_5)(C_2H_4)]^+$. (Figure reproduced with permission from [83])

attack, the transition state search (carried out at the DFT-LDA level with a polarized basis set) led to a structure of C_{2v} symmetry (see Fig. 6). A single imaginary frequency was computed, thus confirming that this structure represents a transition state. However, the corresponding normal mode had the characteristics of a reaction coordinate representing the hydrogen exchange process and not the insertion process. To put this on more firm ground the intrinsic reaction coordinate (IRC) path was studied. We mentioned earlier the high computational cost of this type of calculation. The relatively small size of the model system chosen here allows such a determination which would probably not have been feasible with two cyclopentadienyl ligands. The present example is well representative of the dilemma that we mentioned in the introduction, between the sophistication of a calculation and the relevance of the model system used for this calculation.

The results of the IRC calculations are best illustrated in Fig. 7, which shows the change in the H_1-C_β and $H_1-C_{\beta'}$ distances. One bond distance is gradually broken while the other is formed. This points clearly to the hydrogen exchange process. The corresponding energy barrier amounts to 5.3 kcal mol^{-1}.

All attempts to locate a transition state for the insertion were unsuccessful, leading to the conclusion that the insertion through the frontside attack on the β-agostic conformer is either unlikely, or unable to compete with the hydrogen exchange.

The backside attack was considered next and its transition state located. In agreement with the previous calculation of Morokuma for reaction (1), the geom-

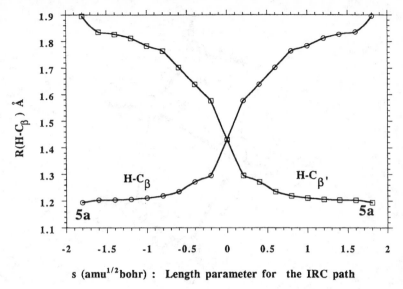

Fig. 7. Change in the $R(C_\beta\text{-}H_1)$ and $R(C_{\beta'}\text{-}H_1)$ distances along the intrinsic reaction coordinate (IRC) of the hydrogen exchange process in $[Cl_2Ti(C_2H_5)(C_2H_4)]^+$. (Figure reproduced with permission from [83])

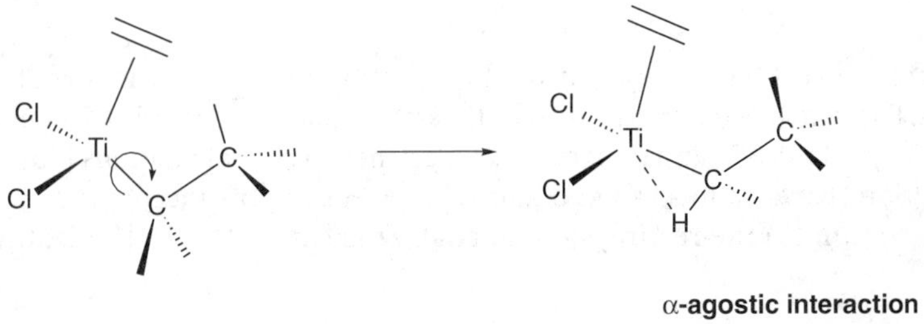

α-agostic interaction

Scheme 6

etry of the transition state departs somewhat from Cs symmetry (see Fig. 8). A transition state structure of Cs symmetry was found, but three imaginary frequencies were obtained from the vibrational analysis, whereas only one was obtained from the vibrational analysis for the transition state of Fig. 8. An IRC calculation again showed that this TS does indeed correspond to the backside insertion [83]. The energy barrier (computed at the DFT-NL level) amounts to 3.9 kcal mol⁻¹.

It can therefore be concluded that the frontside insertion of C_2H_4 into the metal–carbon bond of the *β-agostic* structure is unlikely. At this stage, however, it is important to recall that all previous studies on the insertion of C_2H_4 into the metal–methyl bond stressed the role of the α-agostic interaction in the transition state. One can therefore imagine some rotation of the ethyl ligand around the metal–C_α bond to form an α-agostic intermediate that would then undergo a frontside insertion, see Scheme 6. Reasoning along these lines Ziegler and

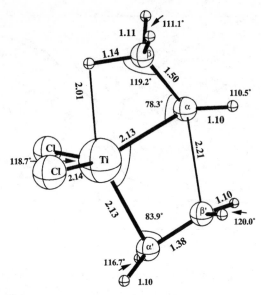

Fig. 8. Optimized (DFT-LDA level) structure of the transition state for the backside insertion of C_2H_4 into the Ti-C_α bond of $[Cl_2Ti(C_2H_5)]^+$. (Figure reproduced with permission from [83])

coworkers investigated very thoroughly reaction (5) which is a more realistic model. Their investigation involved both "static" quantum chemical calculations (of DFT type) [85,86], and dynamic Car–Parrinello type calculations [93]. These calculations have led to a rather detailed knowledge of the mechanism.

In reaction (5) the resting state to analyze is $[Cp_2Zr(C_4H_9)]^+$, viz. the

$$\left[Cp_2Zr(C_2H_5) \right]^+ + C_2H_4 \longrightarrow \left[Cp_2Zr(C_4H_9) \right]^+ \tag{5}$$

butyl derivative. Four possible orientations of the alkyl chain were considered, corresponding to the α-, β-, γ- or δ-H agostic interaction. As expected the most stable conformer is the β-agostic one, whereas the α-agostic conformer is destabilized by 11.2 kcal mol^{-1}. The γ-agostic and δ-agostic structures lie in between, being destabilized by 6.4 and 7.4 kcal mol^{-1}, respectively. Thus the resting state is clearly the β-agostic conformer. However, as in reaction (4), a direct frontside attack of the β-conformer does not lead to insertion chain growth, but to chain termination through H-exchange, which has a relatively low barrier, 6.7 kcal mol^{-1}. Ziegler et al. then considered the rotation around the Zr–Cα bond [86]. The DFT calculations show that the rotation is relatively easy in the β-agostic π-complex obtained after the coordination of ethylene: the barrier amounts to 3.3 kcal mol^{-1} only. This rotation leads to the α-agostic π-complex. The insertion of C_2H_4 in the α-agostic Zr–C_2H_5 conformer is very easy. As for the insertion into Zr–CH_3 there is almost no barrier (0.5 kcal mol^{-1}). Thus it seems clear from these calculations that *the propagation barrier has its origin in the rearrangement of the β-agostic conformer to the α-agostic conformer*. The backside insertion was also analyzed and its transition state determined, with a barrier amounting to 6.8 kcal mol^{-1}.

One should mention here that the geometries of the transition states found for the H-exchange and the backside insertion in reaction (5) are very similar to those obtained previously with the model reaction (4). The transition state for the frontside insertion in the α-agostic structure (Fig. 9) shows clearly the involvement of the agostic interaction. The difference between the energy barriers computed for the chain termination via the H-exchange reaction (6.7 kcal mol^{-1}) and for the propagation step (3.3 kcal mol^{-1}) does not account, when applying Maxwell–Boltzmann statistics, for the molecular weight obtained experimentally. These values, however, correspond to "electronic" energy barriers $\Delta E^{\neq}$ and not to free energy barriers $\Delta G^{\neq}$ since they do not take into account the differences in zero-point energy, in entropy and in vibrational energy. These quantities can be obtained from a thermal analysis of the reactants and transition states. When this is carried out, a negligible free energy of activation for the rotation around the Zr–C_α bond ($\Delta G^{\neq} < 0.1$ kcal mol^{-1}) is obtained. The value obtained for the H-exchange process is much greater, 5.8 kcal mol^{-1}. The difference between these two values corresponds better to the molecular weight obtained experimentally. The free energy of activation computed for the backside insertion is also greater that for the rotation of the ethyl group, 6.4 kcal mol^{-1}, making this reaction quite unlikely. A detailed analysis of the various contributions revealed

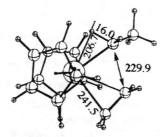

Fig. 9. Optimized (DFT-LDA level) structure of the transition state for the frontside insertion of C_2H_4 into the Ti–C_α bond in the α-agostic π-complex [$Cp_2Zr(C_2H_5)(C_2H_4)$]$^+$. (Figure reproduced with permission from [86])

that an important factor (reflected in the positive $\Delta S^{\neq}$ and a large change in the zero-point energy) is the high mobility of the ethyl group.

The above findings point to the importance of the dynamic effects. To assess more precisely these effects, Ziegler and coworkers carried out dynamic calculations on the same reaction [93]. They used the technique based on the constrained Car–Parrinello molecular dynamics that we mentioned in Sect. 2.4. The simulation (done at 300°K) confirmed the static picture. In particular, the agostic interactions were found to shift rapidly, in line with the high mobility of the ethyl group deduced from the static calculations. Figure 10 [93] displays the behaviour of the force $\delta E/\delta\lambda$ on the reaction coordinate (shaded line) and its integral. The arrows indicate the location of the DFT stationary points (precursor starting structure FS-S, transition state FS-T and γ-agostic end product FS-E). The good agreement between the maximum and minimum of the integral curve and the location of the DFT optimized transition state and products, respectively, is quite noteworthy. The total simulation time was not long enough to obtain accurate values of free energy differences, although the value computed from the dynamic simulation (4.8 kcal mol^{-1}) is in qualitative agreement with the value of 2.0 kcal mol^{-1} obtained from the static calculations (after deletion of the quantum vibrational effects that are not taken into account in the dynamic simulation).

Most recently Ziegler et al. have also carried out a similar static and constrained molecular dynamics study of the polymerization of olefins catalyzed by another complex, the so-called constrained geometry catalyst (CGC) of the type $[(CpSiH_2NH)TiR]^+$, see **4**, including the study of the chain termination and chain branching steps [94,95]. With this catalyst the barrier for the propagation

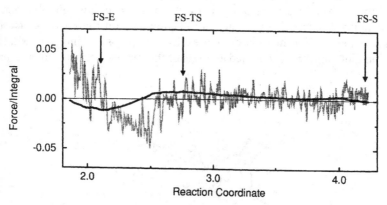

Fig. 10. Force on the reaction coordinate (*shaded line*) and its integral (*full line*). Arrows indicate the location of the FS insertion precursor complex $[Cp_2Zr(C_2H_5)(C_2H_4)]^+$ (FS-S), the FS insertion transition state (FS-TS) and the FS insertion γ-agostic product (FS-E), as determined by the DFT-NLA calculations. Distances are in Å, and the force on the reaction coordinate and its integral are in atomic units (1 a.u.=627.5 kcal mol^{-1}). (Figure reproduced with permission from [86])

Structure 4

step is found to have its origin in the olefin insertion into the γ-agostic structure. The barrier is also higher than in reaction (5). Interestingly, a full molecular dynamics study was attempted by Meier et al. [92]. The simulation showed no sign of the insertion taking place within the time scale of the simulation, 900 fs. Instead, other events were observed, such as the rotation of C_2H_4 away from the configuration with the ethylene C=C bond parallel to the metal R bond. This result best illustrates the discussion made in Sect. 2.4 about the advantage of performing constrained CP dynamics when there is a non-negligible reaction barrier.

Finally, to conclude this section, it is worth emphasizing that this accumulation of theoretical knowledge and of the corresponding data has led to a point where comprehensive and rather accurate theoretical studies of the ethylene polymerization by d^0 and $d^0 f^n$ transition-metal complexes with a variety of ligands can now be envisioned. Such studies are currently under way [89,115,116].

3.2
Olefin Polymerization Using Late Transition-Metal Complexes as Catalysts

It is not surprising, when considering the amount of theoretical work that has been devoted to the olefin polymerization catalyzed by early transition-metal complexes, that the recent discovery by Brookhart and coworkers [117–119] of a similar process catalyzed by Ni(II)- and Pd(II)(diimine)(methyl) complexes has also attracted the attention of theoreticians [62,116,120–126]; for another study of olefin polymerization by Ni(II)(acac) complexes, see [127].

The overall mechanism, as proposed by Brookhart and coworkers, is similar to the mechanism for metallocene catalysts. The chain initialization involves coordination and insertion of ethylene into the metal–methyl bond of the metal diimine methyl complex. Next, chain propagation proceeds by coordination and insertion of ethylene into the M–R bond (R being the growing chain). Chain isomerization involves β-elimination of hydrogen from the growing chain, followed by reattachment to the terminal carbon atom. Finally, the chain termination would also involve the β-hydrogen elimination (assisted by an incoming ethylene) from the growing chain. In this case, however, the eliminated hydrogen would be transferred to the coordinated olefin.

The theoretical studies were primarily aimed at: (i) rationalizing the greater activity of Ni(II) vs. Pd(II) catalysts; (ii) delineating the factors that account for the degree of branched vs linear polymer produced, and how these factors vary according to metal (experimentally Ni(II) catalysts produce more linear polymer than Pd(II)); (iii) determining the dominant chain termination step; and (iv) providing a comparison with metallocene-catalyzed polymerization.

It should be stressed at the onset, that both Figs. 2 and 3 explain why the reactions that have been found with d^0 metallocene complexes are also found with d^8 Ni or Pd complexes. T-shaped Ni (or Pd) diimine alkyl complexes have frontier orbitals that are just like those in Fig. 3. Hence, there is nothing from a symmetry point of view that prevents the approach of ethylene and the reaction of its carbon–carbon bond with the metal–carbon bond. The similarity of the frontier orbitals of the d^0 Cp$_2$MR fragment and of a T-shaped d^8 ML$_3$ fragment just materializes the isolobal analogy [21] between the two fragments and some similarity in the reaction pattern of both fragments is to be expected.

The Ni(II) and Pd(II) complexes that are experimentally used have bulky diimine ligands ArN=C(R)C(R)=NAr (where Ar = 2,6-C$_3$H$_6$(i-Pr)$_2$ and 2,6-C$_3$H$_6$Me$_2$ and R=H and Me). The first studies carried out by Morokuma and coworkers [116,120,121], Siegbahn and coworkers[122,123] and Ziegler and coworkers [125] were restricted to the model diimine ligand HN=CH-CH=NH. Although some differences are found between these studies, the broad picture that emerges from all of them is similar. The insertion barrier is relatively low, although somewhat larger than for metallocene complexes (between 9.9 to 17 kcal mol^{-1} for Ni, depending on the studies). Interestingly, too, for these late transition-metal catalysts, no agostic interaction is involved in the transition state. The barriers are higher for Pd(II) and especially for Pt(II) [116,121–123]. This has been explained by the relative strengths of the metal–olefin bond which, inter alia, depends on the amount of back donation from the d_π-orbital into the π^*-orbital of the olefin. That the activation barrier for the β-elimination process is higher in the Ni(II)-catalyzed process can also be traced to the strength of the metal–olefin bond in the product of β-elimination, namely the metal hydride olefin complex.

A salient feature of the studies that used the model diimine ligand is that the activation barriers computed for the chain termination barriers are of the same order of magnitude or even lower than the activation barrier for the propagation step. The experimental studies suggest that the bulkiness of the ligands ArN=C(R)C(R)=NAr (Ar being either 2,6-C$_3$H$_6$(i-Pr)$_2$ or 2,6-C$_3$H$_6$Me$_2$) is a key feature for the polymerization control. More specifically, it has been proposed [117–119] that the bulky aryl ligand would block the axial site of the metal center. Since the calculation of the transition states for chain termination, at least for the olefin-assisted β-elimination, indicate that the axial sites are occupied whereas the insertion transition state corresponding to the chain propagation only involves the equatorial sites, steric effects are likely to be crucial and should be taken into account.

This has been done by Morokuma [124] and by Ziegler [126]. Morokuma and coworkers have relied on their integrated molecular orbital/molecular mechanics method (IMOMM). In this method [128] (see the chapter by Maseras for more details) both the QM (here DFT-B3LYP) part and the MM part (based on the MM3 force field) are optimized simultaneously, thus allowing the relaxation of the QM part (i.e. the active part) to accommodate the bulky ligand. This is at variance with the QM/MM study of the Ziegler–Natta-type reaction (2) (vide supra), where no relaxation of the QM part was allowed [78,90]. Ziegler et al. also used the same IMOMM approach, but with different exchange correlation functionals (local and non-local) in the QM part, and the AMBER 95 force field instead of the MM2 force field in the MM part (see the original publications [124,126] for more details). In addition to this "static" QM/MM approach, Ziegler et al. have also used their combined ab initio molecular dynamics Car–Parrinello QM/MM methodology that was mentioned in Sect. 2.4 [62].

The QM/MM static calculations of both groups led to very similar conclusions and hence stress the power of such combined methods (more details about the similarities and differences can be found at the end of the paper by Morokuma [124]). In particular they both point to the destabilization of the diimine-Ni-olefin-alkyl π complex which, in turn, results in a lower barrier for the insertion barrier of the chain. The calculations of Ziegler also incorporate the determination of the transition state for the ethylene-assisted β-H elimination (one of the possible chain termination steps). The QM/MM calculations for the β-H elimination show clearly a large increase in the barrier when going from the model to the experimental system, from 9.7 to 18.8 kcal mol^{-1}. This supports the proposal of Brookhart et al. [117–119]. Moreover, the calculations now account for the high molecular weight polymer that can be obtained with the catalyst, since the chain propagation barrier is lower than the chain termination barrier. The calculations also reveal some interesting electronic effects of the experimental ligand: there is some preference for the aryl ring of the substituent to move away from a perpendicular alignment to the Ni-diimine ring, in order to gain some part of the stabilization expected when the π-systems are coplanar to each other.

The dynamic simulations confirmed these findings [62]. They were performed for the β-elimination process in the $[(2,6\text{-}C_3H_6(i\text{-}Pr)_2)N=C(Me)C(Me)=N(2,6\text{-}C_3H_6(i\text{-}Pr)_2)]Ni(C_2H_4)(C_3H_7)^+$ system (using the combined QM/MM potential) and in the $[HN=C(H)C(H)=NHN]Ni(C_2H_4)(C_3H_7)^+$ model system (using the QM potential). The free energy difference was obtained by thermodynamic integration on a carefully chosen reaction coordinate between the reactant and the transition state (see [62]) and amounts to 14.8 kcal mol^{-1} with the experimental ligand and to only 9.8 kcal mol^{-1} with the model ligand. The value obtained for the $[(2,6\text{-}C_3H_6(i\text{-}Pr)_2)N=C(Me)C(Me)=N(2,6\text{-}C_3H_6(i\text{-}Pr)_2)]Ni(C_2H_4)(C_3H_7)^+$ system is in good agreement with the experimental estimate of about 16 kcal mol^{-1} (see ref 18 in [62]).

Related to the above work are the studies pertaining to the copolymerization of CO and olefin [129–131]. The studies of this catalytic process were restricted to pure QM methods, either at the DFT level [129–131] or at the post-HF level [131]. In their study, Margl and Ziegler investigated the process catalyzed by the $[Pd(H_2PCH=CHPH_2)]^{2+}$ catalytic center as a model of the Pd(Me-DUPHOS) used experimentally [132]. Morokuma et al. used the ethanediimine ligand HN=

C(H)C(H)=NH to model the experimental 1,10-phenathroline ligand used by Brookhart and Rix [133], and examined Ni and Pd complexes of this ligand. We do not want to detail here the results of these theoretical studies. They account nicely for the experimental findings and provide in particular a rationalization of the strictly alternating (especially for Pd) copolymerization pattern. A coupled experimental/theoretical study on the mechanistic aspects of the alternating copolymerization of propene with carbon monoxide, catalyzed by Pd(II) complexes of unsymmetrical phosphine-phosphite ligands, has also appeared recently [134].

3.3
Olefin Hydroformylation and the Wacker Process: Influence of the Solvent Effects

The theoretical studies that we have analyzed so far do not take into account the reaction medium. As we mentioned in the introduction, solvent effects can be rather important for the course of an organometallic reaction, not only because of the dielectric properties of the solvent but also because of its coordinating properties or hydrogen-bonding capacities. There are no methods which would be peculiar to organometallic chemistry. Instead the methods that have been used in organometallic chemistry have been used previously in organic chemistry.

In order to treat solvent effects two main approaches have been considered. In the microscopic approach both the organometallic system and the surrounding solvent molecules are treated explicitly by quantum mechanical methods, as a "supermolecule". This approach can be useful, especially for determining specific solvation sites in the case of processes taking place in coordinating solvents and involving unsaturated species. Ligand dissociation processes belong to this class. The first solvation sphere also has to be treated using this approach when hydrogen bonds arising from protonic solvents are expected to be important in the reaction under study.

A major drawback of this approach is its computational cost, since the size of the system with its surrounding molecules can be quite large if a large fraction of the solvation energy is to be obtained, or if many molecules contribute to the hydrogen-bonding network around the reacting center. Also to be taken into account are the solvent rearrangement phenomena through the optimization of the solvent coordinates. This is computationally very expensive. Nonetheless, some attempts have been made along these lines as will be shown in some examples.

An alternative approach – the macroscopic one – is to consider the system inside a cavity immersed in a homogeneous solvent, and to represent the solvent as a polarizable and dielectric continuum [135]. The system is considered as a charge distribution. This charge distribution interacts with the solvent and polarizes it. This gives rise to a reaction field that, in turn, interacts with the charge distribution of the system, thus leading to electrostatic stabilization. The reaction field depends on the dielectric constant of the medium ε, and on the shape and on the size of the cavity. The cavity can be spherical, ellipsoidal, or of molecular shape. It should be stressed here that the shape and the size of the cavity may be quite critical [135,136]: if the cavity is too large the solvation effects will be too small. Conversely, if it is too small it may give spurious effects. If its shape is inadequate, this may lead to some distortion.

There has been a whole wealth of methods that have been designed to solve the quantum mechanical treatment – either at the ab initio level or at the semi-empirical level – of a system embedded in the solvent reaction field. The interested reader will find a thorough discussion and classification of these methods in the review article by Tomasi [135]. Among them, special attention should be paid to the so-called self-consistent reaction field (SCRF) methods where the charge distribution of the solute is expressed as a multipole expansion, and an iterative solution is found for a consistent solute charge distribution and solvent reaction field. The calculations can be carried out either at the HF, MPn or DFT level. They are relatively straightforward and computationally not too expensive and, for these reasons, they are therefore the ones which are the most frequently used in the field of organometallic chemistry.

As far as ab initio methods are concerned other methods are also available. The polarized continuum model (PCM) of Tomasi [137,138] uses apparent surface charges (that depend on the dielectric constant ε) with a cavity of molecular shape [139]. It also involves a self-consistent procedure in the calculation. The molecular shape surface is based on van der Waals spheres placed at the atoms. Some variants exist, e.g. the isodensity polarizable continuum model (IPCM) in which the cavity is based on an isosurface of the total electron density calculated at the level of the theory being applied [140,141], or the self-consistent isodensity PCM model (SCI-PCM), a fully self-consistent IPCM method in which the isodensity surface is relaxed during the iterative procedure [141,142].

The continuum model lacks specific information on the intermolecular interactions involved between the solvent and the solute. It cannot treat, for instance, the dative bonds made by a coordinating solvent. Moreover, due to the cavity approximation, reactions extending over a large range of distances cannot be treated. One way to circumvent these shortcomings is to combine a microscopic approach for the first solvation sphere (or for one or several specific interactions) with a macroscopic approach for the remaining effects that are mostly due to the dielectric bulk [143,144]. The "spectator" solvent molecules could also be treated with one electron terms that are added to the Hamiltonian that describes the solute (and eventually the molecules that are explicitly taken into account) [145,146]. Another approach that is gaining increasing attention in organic chemistry is to combine QM/MM methods in Monte Carlo (MC) or molecular dynamics (MD) simulations [147–151]. The solute molecule (and eventually the nearby solvent molecules) are treated at the QM level whereas the solvent molecules (or the more remote ones) in the MM region are approximated by empirical molecular mechanics force fields. MC or MD simulations then lead to the determination of free energies. Since the QM calculations have to be repeated for nearly every MC solute move or MD time step in the MC or MD simulations, this method is, from a computational point of view, extremely expensive and currently impractical for organometallic systems. A promising approach might be to use a combined SCRF/MC calculation [152]. In this method solvent adapted geometries and charges from an ab initio SCRF calculation are obtained and used in MC simulations that include explicit solvent molecules.

The effect of the coordinative properties of a solvent in the course of a catalytic process has been analyzed by Morokuma and coworkers [153] in a theoretical study of the full cycle of the olefin hydroformylation catalyzed by Rh(I) complexes, reaction (6).

$$RCH=CH_2 + CO + H_2 \xrightarrow{\text{RhH(CO)}_2(\text{PR}_3)_2} RCH_2CH_2CHO \tag{6}$$

Some steps of the olefin hydroformylation catalyzed by various catalysts have been theoretically investigated previously [154–157], but the study of Morokuma is the first comprehensive one for this process. The active catalyst was modelled by the $RhH(CO)_2(PH_3)_2$ complex and the alkene-type solvent by an extra C_2H_4 molecule. The inclusion of the solvent effects changes drastically the shape of the free energy profile (obtained once the zero-point energy and entropy contribution have been taken into account through a thermal analysis of the various intermediates and transition states). As expected, all four-coordinate intermediates that are coordinatively unsaturated were found to be strongly stabilized (by more than 20 kcal mol^{-1} at the MP2 level). Their square planar structures are transformed to five-coordinate trigonal bipyramid (TBP) structures, the alkene solvent always lying in the equatorial plane. This preference is best rationalized on the basis of MO arguments that were given some twenty years ago by Rossi and Hoffmann [158]. On the other hand, since all transition states (for the C_2H_4 insertion, the CO insertion, the H_2 oxidative addition and the reductive elimination steps) are five-coordinate complexes, with either a square pyramid or a TBP structure, they do not interact strongly with the solvent ethylene molecule: the Rh–ethylene distances are ~5 Å and the stabilization energy 1.5–3.5 kcal mol^{-1} at the MP2 level. Thus the addition of the entropy contribution causes the solvated transition states to be less stable than the unsolvated ones. This suggests that, in terms of the free energy, there is no coordination of the solvent to the transition states. For the same reasons the ligand substitution step should occur via a dissociative mechanism in solution. Finally the inclusion of the solvent effect, by stabilizing the four-coordinate intermediates, and not the transition states, makes the H_2 oxidative step a likely candidate for the rate-determining step. This is in line with the experimental observation that the overall rate is proportional to the pressure of hydrogen.

The various steps of the Wacker process take place in water. This therefore provides an opportunity to study the effect of a protic solvent on the course of a catalytic reaction. This has been done by Siegbahn [159,160]. We have previously reviewed the theoretical studies that pertain to this multistep process [161]. Most of them were devoted to the step involving the nucleophilic attack of the $PdCl_2(H_2O)(C_2H_4)$ intermediate by water [Eq. (7)].

$$
\underset{\underset{Cl}{|}}{\overset{\overset{Cl}{|}}{H_2O - Pd -}} \| \quad + \quad H_2O \quad \longrightarrow \quad \left[\underset{\underset{Cl}{|}}{\overset{\overset{Cl}{|}}{H_2O - Pd}} \diagup^{-OH} \right]^{-} \quad + \quad H^+ \tag{7}
$$

This attack results in the nucleophilic addition of a hydroxyl to the coordinated olefin to yield $PdCl_2(H_2O)(C_2H_4OH)^-$. The early studies were successful in delineating the driving force for the nucleophilic addition (namely the slipping of the ethylene from η^2 towards η^1 coordination mode), but the energetics obtained from the gas phase model are not realistic. The exothermicity computed by Siegbahn amounts to 77.8 kcal mol^{-1} (DFT-B3LYP level) [160]. SCRF studies (using a spherical cavity and restricting the multipole expansion to the quadrupole moment) take into account some of the bulk dielectric properties and reduce this value to 54.6 kcal mol^{-1}. This, however, is not enough. As shown by Siegbahn, in order to obtain a realistic model one needs to describe the nucleophile by a chain of water molecules (at least 3) held by hydrogen bonds. This chain can bridge the negative chloride ligand and the point of attack of the olefin. At the SCF level this model yields a slight endothermicity for the nucleophilic addition; however, a reasonable exothermicity (14.5 kcal mol^{-1}) is recovered after inclusion of the long range polarization effects via the SCRF calculation. This study is therefore typical of the important role played by short-range *and* long-range solvent effects.

The second step of the process, the β-elimination reaction from $[PdCl_2(H_2O) (C_2H_4OH)]^-$, has also been examined, both in the gas phase model [Eq. (8)] and in the solvated structure $[PdCl_2(H_2O)(C_2H_4OH)^-][H_7O_3^+]$.

$$
\left[\underset{\underset{Cl}{|}}{\overset{\overset{Cl}{|}}{H_2O - Pd}} \diagup^{-OH} \right]^{-} \quad \longrightarrow \quad \underset{\underset{Cl}{|}}{\overset{\overset{H}{|}}{H_2O - Pd -}} \| \diagdown^{OH} \quad + \quad Cl^- \tag{8}
$$

A very high barrier (28 kcal mol^{-1}) is found for the gas phase model, inconsistent with the experimental evidence that points to an easy β-elimination step. In the solvated structure (and also taking into account the bulk properties through a SCRF calculation) the barrier is reduced by about 10 kcal mol^{-1}. Interestingly, the structure of the corresponding transition state shows that one of the chloride ligands moves out of the coordination sphere as the H-transfer to Pd proceeds. It also shows that this movement is assisted by the water molecules sited around the Pd center. A further decrease in the barrier is obtained by enlarging the basis set. It is not sufficient however to reconcile theory and experiment at this point and further studies are needed.

A word of caution about the SCRF treatment is appropriate at this stage. As noted already by Siegbahn [160], large changes of cavity radii should be avoided when dealing with the comparison of solvation energies. Some spurious effects were also noticed by Siegbahn in his SCRF calculations where the length of the multipole expansion was increased from quadrupole to hexadecapole [160]. We have recently completed a theoretical study of such effects in the SCRF method, in the case of Pd(II) and Pd(IV) complexes with various environments [162]. We have found similar features and analyzed accordingly the performance of various cavities (including a new one of ellipsoidal shape, where the size of the ellipsoid follows as closely as possible the molecular shape of the system).

4
Conclusion and Perspectives

We have reviewed, using selected examples taken from homogeneous catalysis, the most advanced methods currently being used in quantum chemistry to study organometallic reaction mechanisms. Other examples of catalytic processes that have been thoroughly and successfully studied and that have not been discussed in this review include the dihydroxylation of olefins catalyzed by OsO_4 [163–168] (for which there is still a controversy about the mechanism of the cycloaddition step of the double bond, either [3+2] or [2+2]) [169]; the hydrogenation of CO_2 into formic acid catalyzed by hydrido bisphosphine rhodium(I) complexes, for which we have proposed an alternative mechanism based on a σ-bond metathesis pathway [170,171]; the hydrogenation of CO catalyzed by $HRh(CO)_4$ [172]; the rhodium(I)-catalyzed olefin hydroboration [173]; the Pt(0)-catalyzed alkyne diboration [174]; and the silastannation of acetylene [175].

It seems clear that the evolution of theoretical organometallic chemistry into *computational* organometallic chemistry and catalysis is irreversible. Studies that have been carried out over the past few years have brought the field to a relatively mature stage. It is now known in most instances which methodology will be appropriate to tackle a given problem, at least in the gas phase, and for first to third row transition metal complexes. As a result, the dialogue with the experimentalist is put on relatively solid ground.

This does not imply that for studying organometallic reaction mechanisms the available softwares can be used as a "black box". There are still cases where a careful cross checking of the various methodologies at hand is needed, in order to assess the validity of the results, or to improve their accuracy. A striking example is provided by the recent studies on the oxidative addition of methane to CpM(CO), M= Co, Rh, Ir [176–179], where the DFT-B3LYP results deviate somewhat from the results obtained at the highest post-HF levels, especially for Co and Rh.

Standard methodologies cannot be used either when dealing with photochemical organometallic reactions. Here, MCSCF/CI techniques must be used to correctly assign the excited states and construct potential energy surfaces. The dynamics can then be simulated by using a time-dependent wave-packet propagation technique [180–182].

Among the topics that should experience new developments and progress in the near future are the QM/MM type approaches and the inclusion of solvent effects. Clearly such developments are needed for reactions in solution and for processes where steric effects are important, e.g. enantioselective catalysis. The inclusion of dynamics, not only in the gas phase but also in solution [183,184], should very soon be an active field of research.

All these development will pave the way for new applications in homogeneous catalysis, in biochemistry, e.g. enzymatic reactions in which the active center involves one or several metals where studies on model reactions have already been carried out [185–188]. More investigations on reactivity in lanthanide and actinide chemistry [48,189] can also be expected. The near future will certainly see more and more *integrated* experimental and theoretical studies, clarifying or even predicting reaction mechanisms in these fields.

References

1. Koga N, Morokuma K (1991) Chem Rev 91:823
2. Ziegler T (1991) Chem Rev 91:651
3. Ziegler T (1995) Can J Chem 73:743
4. Dedieu A (ed) (1992) Transition metal hydrides. VCH, New York
5. Yoshida S, Sakaki S, Kobayashi H (1994) Electronic processes in catalysis. VCH, New York
6. Musaev DG, Matsubara T, Mebel AM, Koga N, Morokuma K (1995) Pure Appl Chem 67:257
7. Musaev DG, Morokuma K (1996) In: Prigogine I, Rice SA (eds) Advances in chemical physics. Wiley, New York, Vol. XCV, p 61
8. Simons J (1991) J Phys Chem 95:1017
9. Salahub DR, Zerner M (eds) (1991) Metal-ligand interactions: from atoms, to clusters, to surfaces, Kluwer Dordrecht, The Netherlands
10. van Leeuwen PWNM, van Lenthe JH, Morokuma K (eds) (1994) Theoretical aspects of homogeneous catalysis, applications of ab initio molecular orbital theory, Kluwer Dordrecht, The Netherlands
11. Veillard A (1991) Chem Rev 91:743
12. Bauschlicher CW, Langhoff SR, Partridge, H (1995) In: Yarkony DR (ed) Modern electronic structure theory. World Scientific, London
13. Frenking G, Antes I, Böhme M, Dapprich S, Ehlers AW, Jonas J, Neuhaus A, Otto M, Stegmann R, Veldkamp A, Vyboishchikov SF (1996) In: Lipkowitz KB, Boyd DB (eds) Reviews in computational chemistry. VCH, New York, vol 8, p 63
14. Siegbahn PEM (1996) In: Prigogine I, Rice SA (eds) Advances in chemical physics. Wiley, New York, vol XCIII, p 333
15. Hoffmann R (1963) J Chem Phys 39:1397
16. Woodward RB, Hoffmann R (1970) The conservation of orbital symmetry, Verlag Chemie, Weinheim, Germany
17. Pearson RG (1976) Symmetry rules for chemical reactions. Wiley, New York
18. Albright TA (1985) Tetrahedron 38:1339
19. Albright TA, Burdett JK, Whangbo MH (1985) Orbital interactions in chemistry. Wiley, New York
20. Hoffmann R (1981) Science 211:995
21. Hoffmann R (1982) Angew Chem Int Ed Engl 21:711
22. Whangbo MH, Schlegel BH, Wolfe S (1977) J Am Chem Soc 97:1296
23. Allón C, Alvarez S (1996) Inorg Chem 35:3137
24. Salem L (1982) Electrons in chemical reactions: first principles. Wiley, New York
25. Turro NJ (1978) Modern molecular photochemistry. Benjamin, Menlo Park, California

26. Schlegel HB (1995) In: Modern electronic structure Theory, Part I. World Scientific, London, p 459
27. Müller K (1980) Angew Chem Int Ed Engl 19:1
28. Pulay P (1995) In: Modern electronic structure theory, Part II. World Scientific, London, p 1191
29. Pulay P(1987) Adv Chem Phys 69:241
30. Jørgensen P, Simons J (eds) (1986) Geometrical derivatives of energy surfaces and molecular properties. Reidel, Dordrecht
31. Helgaker T, Jørgensen P (1988) Adv Quant Chem 19:183
32. Fukui K (1981) Acc Chem Res 14:363
33. Gonzales C, Schlegel HB (1994) J Phys Chem 94:5523
34. Gonzales C, Schlegel HB (1989) J Chem Phys 90:2154
35. Ishida K, Morokuma K, Komornicki A (1977) J Chem Phys 66:2153
36. Espinosa-Garcia J, Corchardo JC (1995) J Phys Chem 99:8613
37. Bartlett RJ, Stanton JF (1994) In: Lipkowitz KB, Boyd DB (eds) Reviews in computational chemistry, VCH, New York, p 65
38. Boys SF (1950) Proc Roy Soc London A201:125
39. Roos BO, Taylor PR, Siegbahn PEM (1980) Chem Phys 48:157
40. Siegbahn PEM, Almlöf J, Heiberg A, Roos BO (1981) J Chem Phys 74:2384
41. Roos BO (1980) Int J Quant Chem 14:175
42. Møller C, Plesset MS (1934) Phys Rev 46:618
43. Bartlett RJ, Silver DM (1975) J Chem Phys 62:3258
44. Pople JA, Krishnan R, Schlegel HB (1978) Int J Quant Chem 14:545
45. Andersson K, Malmqvist PÅ, Roos BO (1992) J Chem Phys 96:1218
46. Cizek J (1969) Adv Chem Phys 14:35
47. Pople JA, Head-Gordon M, Raghavachari K (1987) J Chem Phys 87:5968
48. Gordon MS, Cundari, TR (1996) Coord Chem Rev 147:87
49. Christiansen PA, Ermler WC, Pitzer KS (1985) Ann Rev Phys Chem 36:407
50. Pyykkö P (1988) Chem Rev 88:563
51. Kohn W, Sham LJ (1965) Phys Rev A140:1133
52. Parr RG, Yang W (1989) Density functional theory of atoms and molecules. Oxford University Press, Oxford
53. Seminario JM, Politzer P (eds) (1995) Modern density functional theory: a tool for chemistry. Elsevier, Amsterdam
54. Ernzerhof M, Perdew JP, Burke K (1996) Top Curr Chem 180:2
55. Becke AD (1988) Phys Rev A38:3098
56. Lee C, Yang W, Parr R (1988) Phys Rev B37:785
57. Becke AD (1988) Phys Rev A38:3098
58. Car R, Parrinello M (1985) Phys Rev Lett 55:5648
59. Remler DK, Madden PA (1990) Mol Phys 70:921
60. Margl P, Ziegler T, Blöchl PE (1995) J Am Chem Soc 117:12625
61. Kollman P (1993) Chem Rev 93:2395
62. Woo TK, Margl PM, Blöchl PE, Ziegler T (1997) J Chem Phys B 101:7877
63. Brintzinger HH, Fischer D, Mülhaupt R, Rieger B, Waymouth RM (1995) Angew Chem Int Ed Engl 34:1143
64. Thayer AM (1995) Chem Eng News, Sept 11, p 15
65. Cossee P (1964) J Catal 3:80
66. Arlman EJ, Cossee P (1964) J Catal 3:9
67. Brookhart M, Green MLH (1983) J Organomet Chem 250:395
68. Lauher JW, Hoffmann R (1976) J Am Chem Soc. 98:1729
69. Jolly CA, Marynick DS (1989) J Am Chem Soc 111:7968
70. Hyla-Kryspin I, Niu S, Gleiter R (1995) Organometallics 14:964
71. Tatsumi K, Nakamura A, Hofmann P, Stauffert P, Hoffmann R (1985) J Am Chem Soc 107:4440

72. Fukui H (1971) Acc Chem Res 4:57
73. Prosenc MH, Janiak C, Brintzinger HH (1992) Organometallics 11:4036
74. Janiak C (1993) J Organomet Chem 452:63
75. Mohr R, Berke H, Erker G (1993) Helv Chim Acta 76:1389
76. Fujimoto H, Yamasaki T, Mizutani H, Koga N (1985) J Am Chem Soc 107:6157
77. Kawamura-Kuribayashi H, Koga N, Morokuma K (1992) J Am Chem Soc 114:2359
78. Kawamura-Kuribayashi H, Koga N, Morokuma K (1992) J Am Chem Soc 114:8687
79. Siegbahn PEM (1993) Chem Phys Lett 205:290
80. Weiss H, Ehrig M, Ahlrichs R (1994) J Am Chem Soc 116:4919
81. Bierwagen EP, Bercaw JE, Goddard WA III (1994) J Am Chem Soc 116:1481
82. Woo TK, Fan L, Ziegler T (1994) Organometallics 13:432
83. Fan L, Harrison D, Deng L, Woo TK, Swerhone D, Ziegler T (1994) Can J Chem 73:989
84. Woo TK, Fan L, Ziegler T, (1994) Organometallics 13:2252
85. Lohrenz JW, Woo TK, Fan L, Ziegler T (1995) J Organomet Chem 497:91
86. Lohrenz JCW, Woo TK, Ziegler T (1995) J Am Chem Soc 117:12793
87. Yoshida T, Koga N, Morokuma K (1995) Organometallics 14:746
88. Fan L, Harrison D, Woo TK, Ziegler T (1995) Organometallics 14:2018
89. Margl PM, Deng L, Ziegler T (1998) Organometallics 17:933
90. Yoshida T, Koga N, Morokuma K (1996) Organometallics 15:766
91. Meier RJ, van Doremaele GHJ, Iarlori S, Buda F (1994) J Am Chem Soc 116:7274
92. Iarlori S, Buda F, Meier RJ, Van Doremaele GHJ (1996) Mol Phys 87:801
93. Margl P, Lohrenz JCW, Ziegler T, Blöchl PE (1996) J Am Chem Soc 118:4434
94. Woo TK, Margl PM, Lohrenz JCW, Blöchl PE, Ziegler T (1996) J Am Chem Soc 118:13021
95. Woo TK, Margl PM, Ziegler T, Blöchl PE (1997) Organometallics 16:3454
96. Fujimoto H, Koga N, Fukui k (1981) J Am Chem Soc 103:7452
97. Chien JCW, Razavi A (1988) J Polym Sci Part A: Polym Chem 26:2369
98. Chien JCW, Bor-Ping Wang (1990) J Polym Sci Part A: Polym Chem 28:15
99. Chien JCW, Tasai W-M, Rausch MD (1991) J Am Chem Soc 113:8570
100. Chien JCW, Sugimoto R (1991) J. Polym Sci Part A: Polym Chem 29:459
101. Corradini P, Guerra G (1991) Prog Polym Sci 16:239
102. Cavallo L, Guerra G, Vacatello M, Corradini P (1991) Chirality 3:299
103. Cavallo L, Corradini P, Guerra G, Vacatello M (1991) Polymer 32:1329
104. Castonguay LA, Rappé AK (1992) J Am Chem Soc 114:5832
105. Cavallo L, Guerra G, Corradini P (1993) Macromolecules 26:260
106. Hart JR, Rappé AK (1993) J Am Chem Soc 115:6159
107. Rappé AK, Goddard III WA (1982) J Am Chem Soc 104:297
108. Rappé AK, Upton, TH (1984) Organometallics 3:1440
109. Upton TH, Rappé AK (1985) J Am Chem Soc 107:1206
110. Weiss H, Haase F, Ahlrichs R (1992) Chem Phys Lett 194:492
111. Axe FU, Coffin JM (1994) J Phys Chem 98:2567
112. Burger BJ, Thompson ME, Cotter WD, Bercaw JE (1990) J Am Chem Soc 112:566
113. Thompson ME, Buxter SM, Bulls AR, Burger BJ, Nolan MC, Santarsiero BD, Schaefer-WP, Bercaw JE (1987) J Am Chem Soc 109:203
114. Sini G, Mcgregor SA, Eisenstein O, Teuben JH (1994) Organometallics 13:1049
115. Froese RD, Musaev DG, Matsubara T, Morokuma K (1997) J Am Chem Soc 119:7190
116. Musaev DG,Froese RDJ, Morokuma K (1997) New J Chem 21:1269
117. Johnson, LK, Killian CM, Brookhart M (1995) J Am Chem Soc 117:6414
118. Johnson LK, Mecking S, Brookhart M (1996) J Am Chem Soc 118:267
119. Killian CM, Tempel DT, Johnson LK, Brookhart M (1996) J Am Chem Soc 118:11664
120. Musaev DJ, Froese RDJ, Svensson M, Morokuma K (1997) J Am Chem Soc 119:369
121. Musaev DJ, Svensson M, Morokuma K, Strömberg S, Zetterberg K, Siegbahn PEM (1997) Organometallics 16:1933

122. Siegbahn PEM, Strömberg S, Zetterberg K (1996) Organometallics 15:5542
123. Strömberg S, Zetterberg K, Siegbahn PEM (1997) J Chem Soc Dalton Trans 4147
124. Froese RDJ, Musaev DG, Morokuma K (1998) J Am Chem Soc 120:1581
125. Deng L, Margl P, Ziegler T (1997) J Am Chem Soc 119:1094
126. Deng L, Woo TK, Cavallo L, Margl PM, Ziegler T (1997) J Am Chem Soc 119:6177
127. Fan L, Krzywicki A, Somogyvari A, Ziegler T (1996) Inorg Chem 35:4003
128. Maseras F, Morokuma K (1995) J Comput Chem 16:1170
129. Margl P, Ziegler T (1996) J Am Chem Soc 118:7337
130. Margl P, Ziegler T (1996) Organometallics 15:5519
131. Svensson M, Matsubara T, Morokuma K (1996) Organometallics 15:5558
132. Jiang Z, Sen A (1995) J Am Chem Soc 117:4455
133. Rix CF, Brookhart M (1995) J Am Chem Soc 117:1137
134. Nozaki K, Sato N, Tonomura Y, Yasutomi M, Tayaka H, Hiyamaya T, Matsubara T, Koga
     N (1997) J Am Chem Soc 119:12779
135. Tomasi J, Persico M (1994) Chem Rev 94:2027
136. Ford G, Wang B (1992) J Comput Chem 13:229
137. Miertus S, Scrocco E, Tomasi J (1981) J Chem Phys 55:117
138. Tomasi J, Bonaccorsi R, Cammi R, Olivares del Valle JF (1991) J Mol Struct (Theochem)
     234:401
139. Cammi R, Tomasi J (1995) J Comput Chem 16:1449
140. Wiberg KB, Rablen PR, Rusch DJ, Keith TA (1995) J Am Chem Soc 117:4261
141. Foresman JB, Keuth TA, Wiberg KB, Sboonian J, Frisch MJ (1996) J Am Chem Soc
     100:16098
142. Wiberg KB, Castyon H, Keith TA (1996) J Comput Chem 17:185
143. Warshel A, Chu ZT (1994) ACS Symp Ser 568:71
144. Siegbahn PEM, Crabtree RH (1996) Mol Phys 89:279
145. Day PN, Jensen JH, Gordon MS, Webb SP, Stevens WJ, Krauss M, Garmer D, Basch H,
     Cohen D (1996) J Chem Phys 105:1968
146. Chen W, Gordon MS (1996) J Chem Phys 105:11081
147. Åqvist J, Warshel A (1993) Chem Rev 93:2523
148. Field MJ, Bash PA, Karplus M (1990) J Comput Chem 11:700
149. Gao J, Xia X (1992) Science 258:631
150. Gao J (1996) Acc Chem Res 29:298
151. Hartsough DS, Merz KM (1995) J Phys Chem 99:11266
152. Lim D, Jorgensen WL (1996) J Phys Chem 100:17490
153. Matsubara T, Koga N, Ding Y, Musaev DJ, Morokuma K (1997) Organometallics
     16:1065
154. Versluis L, Ziegler T, Baerends EJ, Ravenek W (1989) J Am Chem Soc 111:2018
155. Versluis L, Ziegler T (1990) Organometallics 9:2985
156. Koga N, Jin SQ, Morokuma K (1988) J Am Chem Soc 110:3417
157. Schmid R, Herrmann WA Frenking G (1997) Organometallics 16:701
158. Rossi A, Hoffmann R (1975) Inorg Chem 14:365
159. Siegbahn PEM (1995) J Am Chem Soc 117:5409
160. Siegbahn PEM (1996) J Phys Chem 100:14672
161. Dedieu A (1994) In: van Leeuwen, PWNM, van Lenthe JH, Morokuma K (eds) (1994)
     Theoretical aspects of homogeneous catalysis, applications of ab initio molecular or-
     bital theory, Kluwer Dordrecht, The Netherlands, p 167
162. Visentin T, Dedieu A, Kochanski E, Padel L (1999) Theochem 459:201
163. Veldkamp A, Frenking G (1994) J Am Chem Soc 116:4937
164. Dapprich S, Ujaque G, Maseras F, Lledos A, Musaev DG, Morokuma K (1996) J Am
     Chem Soc 118:11660
165. Pidun U, Boehme C, Frenking G (1996) Angew Chem Int Ed Engl 35:2817
166. Torrent M, Deng L, Duran M, Sola M, Ziegler T (1997) Organometallics 16:13

167. Del Monte AJ, Haller J, Houk KN, Sharpless KB, Singleton DA, Strassner T, Thomas AA (1997) J Am Chem Soc 119:9907
168. Haller J, Strassner T, Houk KN (1997) J Am Chem Soc 119:8031
169. Rouhi AM (1997) Chem Eng News, November 3
170. Hutschka F, Dedieu A, Eichberger M, Fornika R, Leitner W (1997) J Am Chem Soc 119:4432
171. Hutschka F, Dedieu A (1997) J Chem Soc Dalton Trans 1899
172. Pidun U, Frenking G (1998) Chem Eur J 4:522
173. Musaev DG, Mebel AM, Morokuma K (1994) J Am Chem Soc 116:10693
174. Cui Q, Musaev DG, Morokuma K (1997) Organometallics 16:1355
175. Hada M, Tanaka Y, Ito M, Murakami M, Amii H, Ito Y, Nakatsuji H (1994) J Am Chem Soc 116:8754
176. Song J, Hall MB (1993) Organometallics 12:3118
177. Musaev DG, Morokuma K (1995) J Am Chem Soc 117:799
178. Siegbahn PEM (1996) J Am Chem Soc 118:1487
179. Couty M, Bayse CA, Jiménez-Cataño R, Hall MB (1996) J Phys Chem 100:13978
180. Daniel C, Heitz MC, Lehr L, Schröder T, Warmuth B (1994) Int J Quant Chem 52:71
181. Heitz MC, Finger K, Daniel C (1997) Coord Chem Rev 159:171
182. Guillaumont D, Finger K, Hachey MR, Daniel C (1998) Coord Chem Rev, in press
183. Muruguma C, Koga N, Kitaura K, Morokuma K (1994) Chem Phys Lett 224:139
184. Muruguma C, Koga N, Kitaura K, Morokuma K (1995) J Chem Phys 103:9274
185. Pietsch MA, Hall MB (1996) Inorg Chem 35:1273
186. Siegbahn PEM, Crabtree RH (1997) J Am Chem Soc 119:3103
187. Blomberg MRA, Siegbahn PEM, Styring S, Babcock GT, Åkermark B, Korall P (1997) J Am Chem Soc 119:8285
188. Pavlov M, Siegbahn PEM, Blomberg MRA, Crabtree RH (1998) J Am Chem Soc 120:548
189. Di Bella S, Lanza G, Fragalà IL, Marks TJ (1996) Organometallics 15:3986

A Critical Assessment of Density Functional Theory with Regard to Applications in Organometallic Chemistry

A. Görling, S.B. Trickey, P. Gisdakis, N. Rösch*

A. Görling (e-mail: goerling@ch.tum.de), P. Gisdakis (e-mail: gisdakis@ch.tum.de), N. Rösch (e-mail: roesch@ch.tum.de)
Lehrstuhl für Theoretische Chemie, Technische Universität München, D-85747 Garching, Germany
S.B. Trickey (e-mail: trickey@qtp.ufl.edu)
Quantum Theory Project, University of Florida, Gainesville, FL 32611–8345, USA

Reliable quantitative predictions from quantum chemical calculations are a rather recent part of the organo-metallic chemistry scene. The delay, compared to the rate of development in the case of small- to medium-sized organic molecules, was caused largely by the wide variety of valence electronic structures which come into play for transition metal organometallics. Such diversity of hybridization candidates is a challenge to commonly used Density Functional approximations, so it has been somewhat surprising that much of the recent theoretical and computational progress has come from DF calculations. For insight into both the power and limitations of current DF methodology, therefore, we give a descriptive, detailed, but minimally mathematical survey of the ingredients of Hohenberg-Kohn-Sham theory as currently practiced. This overview is followed by a description of the techniques used to build realistic chemistry and physics into the required approximations. After that the origins, properties, and inter-relationships of the more widely used approximations are discussed. In the final section, we give a brief account of the accuracy of current exchange-correlation approximations and of the way in which semi-empirical DF variants are calibrated. A suggestive survey highlights applications to transition metal organometallics, and, as a detailed example, a case study of oxygen transfer reactions by transition metal oxo and peroxo complexes is presented.

Keywords. Density functional theory, Kohn-Sham formalism, Exchange-correlation approximation, Local density approximation, Generalized gradient approximation, Hybrid methods, Olefin epoxidation, Olefin dihydroxylation

Topics in Organometallic Chemistry, Vol. 4
Volume Editors: J.M Brown and P. Hofmann
© Springer-Verlag Berlin Heidelberg 1999

List of Abbreviations

AAD	average absolute deviation
AD	average deviation
B3LYP	Becke-"three parameter"-Lee-Yang-Parr (approximation)
BDE	bond dissociation energy
BLYP	Becke-Lee-Yang-Parr (approximation)
BP86, BP91	Becke-Perdew (approximation) 1986, 1991
CCSD(T)	coupled cluster singles and doubles (perturbative triples)
CI	configuration interaction
DF(T)	density functional (theory)
ECP	effective core potential
GGA	generalized gradient approximation
HF	Hartree-Fock (method)
HK	Hohenberg-Kohn (formalism)
HOMO	highest occupied molecular orbital
KS	Kohn-Sham (method)
IR	infrared
LanL2DZ	basis set with double-ζ quality for Los Alamos National Laboratory ECPs
MCPF	modified coupled pair functional

MD	maximum deviation
MO	molecular orbital
MP2	Møller-Plesset perturbation theory (second order)
MRCI	multi-reference configuration interaction
MTO	methyl trioxorhenium
L(S)DA	local (spin) density approximation
LUMO	lowest unoccupied molecular orbital
LYP	Lee-Yang-Parr (approximation)
PCI-80	parametrized CI
PW86, PW91	Perdew-Wang (approximation) 1986, 1991
QCISD(T)	quadratic configuration interaction including singles and doubles (perturbative triples)
RHF	restricted Hartree-Fock (method)
SCF	self-consistent field
SVWN	Slater-Vosko-Wilk-Nusair (local density approximation)
TFD	Thomas-Fermi-Dirac (method)
TM	transition metal
TS	transition state
UHF	unrestricted Hartree-Fock (method)
VWN	Vosko-Wilk-Nusair (approximation)

1
Introduction

1.1
Significant Problems and the Historical Role of Density Functional Theory

The 1970s witnessed the first impact of quantum chemistry on organometallic chemistry, mostly in the form of qualitative molecular orbital (MO) theory [1–3]. Progress with more sophisticated methods was noticeably slower. To a large extent this delay was connected to the fact that transition metal (TM) complexes remained a significant challenge to theory and computation until about ten years ago. Certain progress was made as pseudopotentials [4,5] became a mature part of Hartree-Fock (HF) based first-principles methods. Further progress came with the direct SCF [6,7] and CI techniques [8,9]. Of course, there is organometallic chemistry of sp metals that has stimulated very fruitful computational activities [10]. However, it is fair to say that to a large extent organometallic chemistry owes its diversity and importance to TM compounds. Thus, the difficulties quantum chemistry had with providing a quantitative description of d and f metal compounds presented an inherent obstacle to a computational treatment of organometallic compounds, let alone their reactions.

These specific difficulties are intimately connected to the diverse chemistry of those compounds, which in turn derives from the diversity of the valence electron structure of TM complexes. Compounds of TM atoms and ions admit large variations in their electronic and geometric structure. Stable, electronically saturated, closed shell molecules can be identified with the help of the 18-electron rule, but even in these reference cases one notes many arrangements of ligands with varying electron donating or accepting character. Exceptions to the rule abound in organometallic chemistry, at variance with the situation for organic compounds and the corresponding octet rule. Ultimately, this rich chemistry of TM atoms and ions can be traced to their flexible valence shell electronic structure which admits many configurations within a relatively narrow energy range and thus many more "hybridization" schemes than the main group elements. The formal electron count on the metal, i.e. its oxidation state, can vary widely and those electrons can form a variety of configurations involving nearly degenerate $(n-1)$d and ns shells as well as, to some extent, the np shell of the TM center.

Thus, the variations in the reactivity of these open-shell structures as well as the difficulties in finding a proper quantum chemical description, can be viewed as two sides of one coin. From a theoretical point of view, one has to be prepared to treat a large number of low-lying electronic configurations on the same footing: The simple HF self-consistent field (SCF) procedure based on a one-determinant wave function must be replaced by much more demanding multi-configuration and "complete active space" SCF methods which rely on wave functions involving several "reference" determinants. Unfortunately, taking this so-called "static" electron correlation into account requires substantially higher computational cost compared to HF. At the same time, one has to be prepared to deal with "dynamic" electron correlation in order to achieve a quantitative description of the thermochemistry involved. In quantum chemistry "dynamic" correlation refers to the reshaping of the electron distribution as compared to HF in order to reduce Coulomb repulsion. It leads to lengthy multi-determinant expansions in configuration interaction (CI) treatments. Density Functional Theory (DFT) deals with the matter entirely differently.

These correlation problems are aggravated if reaction paths and activation barriers are to be treated, since transition states are characterized by changes in electron configuration. As an aside we mention that relativistic effects come into play for organometallic compounds of TMs of the second and third row, so additional special measures are required.

Even a superficial glance at recent chemical literature reveals the broad acceptance of computational chemistry as a practical tool for solving complex chemical problems, in particular those involving TM complexes. Not least, this change was brought about by the wide-spread acceptance of computational methods based on DFT. Such methods afford a surprisingly efficient alternative tool for tackling the electron correlation problem, at least for dynamic correlation. One has to be prepared to encounter difficulties when applying present-day formulations of DFT to computations on systems in which near-degenerate configurations are essential, i.e. in certain cases of static correlation.

Curiously enough, it was the early success with hitherto almost prohibitively complex problems of organometallic chemistry, especially obvious in the work of Ziegler and his group, that marked the paradigm change in computational chemistry in the early years of this decade [11–15]. This unexpected progress (at least from the viewpoint of what is commonly called conventional quantum chemistry) through DFT is documented in a number of review articles and conference proceedings [11,16,17].

On the methodological side, the impact of DF methods is rooted in two advances. Energy gradients for automatic geometry optimization [18–20] were introduced in DF codes, which enabled exploitation of the fact that even at the simplest realistic level of theory (Local Spin Density Approximation, LSDA; see Sect. 3.2.1) a surprisingly accurate description of metal–ligand bonds in complexes and their vibrations became possible [11]. However, even more important for the breakthrough of DF methods was the demonstration that more sophisticated exchange-correlation functionals (Generalized Gradient Approximations, GGA; see Sect. 3.2.1) permitted a dramatic reduction of the computational cost of generating accurate thermochemical information [21].

Soon thereafter, with the advent of so-called "hybrid" functionals [22,23] (Sect 3.2.2), an additional step forward in accuracy was accomplished as far as thermochemical information was concerned; applications to metal–carbon bonds were soon to follow. Intense studies have shown in the meantime that DF-based computational chemistry is by now able to compete very successfully with experiment when it comes to obtaining reaction and activation energies, especially for organometallic systems.

1.2
Perspectives and Priorities

Insight in chemistry often comes from interpretation of MOs. There are highly useful MOs in Density Functional Theory (DFT), but they differ, sometimes subtly, from more familiar ones, e.g. those from the Hartree-Fock approximation [24,25]. Thus, we give a slightly unconventional presentation of basic DFT to help focus on the MOs. Note that DFT [26–34] is a relatively young branch of many-body theory, so there are several open questions. Since applications of DFT are our priority in the present context, we have tried to label those issues and present what seems to be the current accepted wisdom.

The presentation is primarily descriptive. Obviously theory means mathematics, but there is good reason to skip the proofs in favor of discussing the structure and implications of the theory. The proofs are readily available anyway; summaries with references to the original literature are found in Refs. [26–28]. More importantly, proofs can seem abstract and uninformative to the non-specialist. So, we begin by detailed consideration of the role of model wave functions. In particular the distinction between Kohn-Sham (KS) and Hartree-Fock (HF) determinants is developed by examination of the Hohenberg-Kohn theorems with respect to the ordinary quantum mechanical variation problem. The main theme is that the KS determinant is uniquely determined by requiring that it produce the exact ground state density, a somewhat indirect criterion which is very different from the energy minimization constraint on the HF determinant. A brief discussion of DFT orbital occupancy and "electron flow" between orbitals of interacting molecular systems follows. Then, motivated by the role of symmetries in organometallic complexes and reactions, we summarize the primary symmetry problems typical of DFT as used today as well as prescriptions for overcoming them. A survey of the essential physical features of the exchange and correlation holes (by which an electron is screened from the others) follows and leads, in turn, to a survey of the various approximate functionals in use, both "pure" and "hybrid"; we also include a short discussion of semi-empirical variants. With this material in hand, we turn to applications. The examples are chosen to illustrate how DFT behaves (and why) and to show the relevance of that behavior for key issues in organometallics. No pretense of a comprehensive review of DF calculations on organometallics is made.

Unless noted, we consider DFT in its "first-principles" form. "First-principles" electronic structure methods are based solely on the molecular Schrödinger equation or its relativistic equivalent, without empirical parameterization. (We use the term "first-principles" rather than "ab initio" because, in computational chemistry, "ab initio" has been appropriated, somewhat misleadingly, to mean solely conventional, i.e. HF- and post HF-based, quantum chemical methods. Note that in computational solid state physics "ab initio" often means a DF calculation!) We emphasize first-principles forms both so that the actual content of the theory can be examined and because much of the currently popular work on semi-empirical DFT is based on calibration studies to data sets (e.g. [21,35]) which while significant for the organic moieties are not relevant to the (transition) metal constituents.

2
Basics of Density Functional Theory – A Descriptive Presentation

2.1
Density Functional Theory of Closed-Shell Systems: Non-Spin-Polarized Theory

2.1.1
Density Functionals and Model Wave Functions

Restricting consideration to clamped nuclei and, for the moment, to non-relativistic treatments, there are two predominant categories of first-principles methods today: the older methods [25], commonly called "conventional", that start (explicitly or implicitly) in the HF framework, and the younger DF methods [26–34].

In the first category, a brute force first-principles molecular electronic structure calculation faces the problem that the molecular Schrödinger equation for N electrons is a coupled second-order differential equation in $3N$ spatial variables. The solutions are complicated wave functions which can be represented as (formally) infinite sums of Slater determinants, each an antisymmetrized product of N orbitals. Direct solution therefore requires prohibitively high computational effort, except for atoms or small diatomics. Worse, high-accuracy approximations, even for the ground and low-lying states, have complicated mathematical structures which are hard to interpret. Tractability and interpretability are, therefore, the goals of schemes that circumvent the explicit N-electron eigenstates, yet yield reliable numerical values.

Initially we will concentrate on the electronic ground state. Though not usually stated as such, both DFT and conventional methods replace the complicated full interacting ground state by simpler model wave functions, hence rely on the following two points: assignment of a unique model wave function to the real molecular ground state and determination of the real ground state properties from the model wave function. However, the criterion to determine the model wave function and the details of its use are quite distinct.

To make the comparisons clear requires a summary of HF-based theory. The starting point, at least in the absence of degeneracies, is a single Slater determinant. For conventional methods this HF determinant serves as model wave function, either explicitly or at least in terms of interpretive concepts. It is uniquely defined by a variational energy criterion: the HF model state is the Slater determinant that yields the lowest energy expectation value of the full (non-relativistic) molecular electronic Hamiltonian. Approximate values of molecular properties, including the total energy, are taken to be the corresponding values for the HF determinant. The MOs in the determinant are eigenstates of a set of effective single-particle Schrödinger equations, the Hartree-Fock equations [24,25]. The difference between the true ground state and HF energies is defined to be the conventional correlation energy [36]. (There is a subtlety regarding this definition in DFT; see below.) Both the correlation energy and corrections to other properties can be approximated systematically by more elaborate subsequent treatments, e.g. Møller-Plesset perturbation theory [37,38] or Configuration Interaction (CI) calculations [38–41].

Today, DFT is put into practice almost exclusively via the KS method [26,29,42]. To the non-specialist, the KS and HF schemes look similar. This superficial resemblance is misleading, so some effort must be spent on pointing out the important differences.

The KS procedure introduces a different model system of N "non-interacting electrons" which has the same electron density (more carefully, electron number density) in its ground state as the real molecular ground state. "Non-interacting electrons" is a technical term to describe the fact that the KS model system Schrödinger equation has no explicit electron–electron interaction term, but rather an effective single-particle potential that, as will be described, accounts for that interaction correctly. Therefore the N-particle KS Schrödinger equation for the model system decouples into a single-particle Schrödinger equation (the single-particle KS equation) and the KS model ground-state wave function is a single Slater determinant of KS orbitals (for nondegenerate ground states). The motivation for introducing this slightly peculiar auxiliary system as a model will become clearer but roughly speaking it is to provide a systematic yet flexible way to generate and manipulate the densities as needed.

The assignment of a non-interacting model wave function is a crucial point that depends on a basic theorem of DFT, the first Hohenberg-Kohn (HK) theorem [26,29,43]. It states that two N-electron systems with different external potentials cannot have the same ground-state electron density. (Of course, potentials which differ only by a constant are excluded.) As a corollary the ground-state electron density uniquely determines the external potential, thus the Hamiltonian, thus, in principle, all the system properties.

More physically, the interpretation of the first HK theorem is this. For a molecule free of external fields, the molecular external potential for the electrons is the nuclear electrostatic potential. Knowledge of the nuclear positions (hence the external potential) and the number of electrons obviously fixes the Hamiltonian and therefore fixes the molecular electronic structure completely. Thus, in DFT language, the molecular electronic structure and properties are functionals of the external potential and electron number. But the ground-state electronic density yields two things: (a) the electron number and (b) by the first HK theorem, the external potential. Therefore the molecular ground-state electronic structure and properties are functionals of the ground-state electron density. In principle the excited states and their properties are also functionals of the ground-state density because they too are determined by the Hamiltonian. Care is necessary at this point: nothing says that the excited state functionals will be the same as the ground-state one. As an aside, a functional is just a function dependent on another function. Here the function in the role of "independent variable" is the density, a three-dimensional spatial function.

Given the first HK theorem, the second HK theorem is unsurprising: the variation principle for the ground state energy is expressible as a variation principle with respect to the density. The minimum occurs at the exact ground-state density and total energy. Therefore, to summarize to here, there is a total energy functional of the density $E[\rho]$ whose minimum is the ground state energy E_0. That minimum occurs at the ground-state density ρ_0. The associated KS model system consists of non-interacting electrons with the property that their ground state is a determinant which gives the exact (real molecule) ground-state density ρ_0.

One may wonder about bothering with the KS determinant and related constructions. After all, the HK theorems guarantee existence of an alternative to the complexity of conventional HF-based approximate solutions of the full Schrödinger equation, via the electron density, a function of three-dimensional space. The obvious route to exploit this guarantee is to find the electron density for the specified external potential and to determine the molecular electronic properties from that density, without introducing a model wave function, MOs, and associated computational costs. Thomas-Fermi-Dirac (TFD) methods [44] follow this path. Unfortunately, even though TFD methods have received much attention, no version so far has yielded results accurate enough to resolve chemical questions.

Sometimes the first HK theorem is said to be important because of the guarantee just stated. The point is more profound, since the external potential also is a function of three spatial dimensions that completely determines the N-electron ground state. The issue raised by the failure of TFD methods to achieve chemical accuracy therefore is this. Is it better to follow the density-based promise of the HK theorems or pursue characterization via the external potential (and the full Schrödinger equation)?

2.1.2
The Kohn-Sham Procedure

The most important answer to date is provided by the KS procedure. This is the case because the HK theorems establish a precise formal relationship between the KS determinant (the model wave function) and the true molecular ground-state wave function. To see this, note that those theorems hold both for the real (interacting) electronic system and the non-interacting model system. Therefore, under the assumption that the ground-state densities of the two systems are the same, the external potential of the model system and its ground state, the KS determinant, are uniquely determined by the true ground-state electron density. Therefore, we can get at ρ_0, which in principle comes from a complicated wave function, by calculations based on a simple wave function, the KS determinant.

A comment to the wary: the requirement that there actually exists a non-interacting system with ground state density ρ_0 is called the v-representability condition [26,28]. It is an unproven but basic assumption made in the KS formalism which has attracted much interest in the fundamental DFT literature. It seems to hold in almost all cases of interest (exceptions include degenerate ground states and may include lanthanide and actinide compounds with open f-shells [45]). The condition is needed because the first HK theorem does not guarantee that every electron density is the ground state density of some electronic system (in particular, of some non-interacting electron system). It also gives no criterion to determine whether an arbitrary electron density is a ground state density.

With the KS model system defined, we compare properties of the KS determinant and the true ground state by considering their respective Schrödinger equations. By definition the KS determinant is the ground state of a non-interacting model Schrödinger equation of the form:

$$[\hat{T}+\hat{v}_s]\Phi_s = E_s\Phi_s \tag{1}$$

where the N-electron kinetic energy operator is

$$\hat{T} = \sum_{i=1,N} (-1/2)\vec{\nabla}_i^2 \tag{2}$$

and the total potential energy

$$\hat{v}_s = \sum_{i=1,N} v_s(\vec{r}_i) \tag{3}$$

arises from a local effective single-particle potential $v_s(\vec{r})$. (Hartree atomic units are used except as noted.) The KS determinant is denoted by Φ_s. [Comment: do not confuse Eq. (1), the N-electron KS equation, with the more common single-particle KS equation, Eq. (4) below.] The effective potential $v_s(\vec{r})$ is determined such that the KS determinant, as required by its definition, yields the exact molecular ground-state density ρ_0; recall v-representability just discussed. Eq. (1) decouples into a one-particle equation for the KS orbitals:

$$[(-1/2)\vec{\nabla}^2 + v_s(\vec{r})]\varphi_i(\vec{r}) = \varepsilon_i \varphi_i(\vec{r}) \tag{4}$$

The N lowest-energy KS eigenvalues add up to the lowest value of E_s, so to get the KS ground state, the orbitals associated with those eigenvalues go into the KS determinant (and are called "occupied"). In common usage, Eq. (4) is simply the KS equation. Already one distinction between DFT and HF theories is evident. The single-particle KS equation is simpler than the HF equations because the former contains only local potentials, unlike the non-local HF exchange operator. Once v_s is known, solution of the KS equation is straightforward.

Next consider the energies given by the KS determinant. The N-particle KS eigenvalue E_s of Eq. (1) is the sum of the occupied KS orbital eigenvalues ε_i. Here is another distinction with respect to HF. E_s is only the expectation value of the model Hamiltonian $\hat{T} + \hat{v}_s$ with respect to the KS determinant. It is not the DFT value for E_0 (nor estimate in the case of approximations) nor the expectation value of the true molecular Hamiltonian with respect to the KS determinant. Fortunately, E_s hardly ever is reported by a computer code, so the risk is of misunderstanding, not misusing a number.

The non-interacting kinetic energy T_s is

$$T_s = \left\langle \Phi_s \middle| \hat{T} \middle| \Phi_s \right\rangle = \sum_{i=1,N} \left\langle \varphi_i \middle| (-1/2)\vec{\nabla}^2 \middle| \varphi_i \right\rangle \tag{5}$$

while the classical Coulomb repulsion energy U is given by

$$U = \frac{1}{2}\int d\vec{r}\,d\vec{r}'\,\frac{\rho(\vec{r})\rho(\vec{r}')}{|\vec{r}-\vec{r}'|} \tag{6}$$

Here the electron (number) density ρ from the KS determinant is

$$\rho(\vec{r}) = \sum_{i=1,N} \varphi_i^*(\vec{r})\varphi_i(\vec{r}) \tag{7}$$

The integral

$$E_{ext} = \int d\vec{r}\,v_{ext}(\vec{r})\rho(\vec{r}) \tag{8}$$

is the interaction energy with the external potential v_{ext}. Because the KS determinant gives the true molecular ground-state density by construction, the values of U calculated from the KS model system and from the true molecular ground state are identical and the same is true for E_{ext}. (This fact contrasts with the HF determinant, for which both the Coulomb repulsion energy as well as the interaction energy with the external potential differ from the corresponding true molecular values.) The KS exchange energy E_x is given by:

$$E_x = \left\langle \Phi_s \middle| \hat{V}_{ee} \middle| \Phi_s \right\rangle - U = -\frac{1}{2} \sum_{i=1,N} \sum_{j=1,N} \sum_{\sigma\sigma'} \int d\vec{r} d\vec{r}' \frac{\varphi_i^*(\vec{r},\sigma)\varphi_j(\vec{r},\sigma)\varphi_j^*(\vec{r}',\sigma')\varphi_i(\vec{r}',\sigma')}{|\vec{r}-\vec{r}'|}$$

(9)

with $\hat{V}_{ee}$ the electron–electron interaction operator

$$\hat{V}_{ee} = \sum_{i=1,N} \sum_{i<j} \frac{1}{|\vec{r}_i - \vec{r}_j|}$$

(10)

and σ is the usual spin index. The KS orbital expressions for the kinetic energy T_s and exchange energy E_x have the same form as the corresponding HF expressions but differ in value because the KS and HF orbitals differ.

Now consider the fully interacting Schrödinger equation for the molecule in its exact ground state Ψ_0 with ground-state energy E_0:

$$\hat{H}\Psi_0 \equiv [\hat{T} + \hat{V}_{ee} + \hat{v}_{ext}]\Psi_0 = E_0\Psi_0$$

(11)

Here the external potential operator includes the potential from the molecular nuclear framework plus any external fields

$$\hat{v}_{ext} = \sum_{i=1,N} v_{ext}(\vec{r}_i)$$

(12)

For normalized Ψ_0, E_0 is of course also the expectation value

$$E_0 = \left\langle \Psi_0 \middle| \hat{H} \middle| \Psi_0 \right\rangle = \left\langle \Psi_0 \middle| \hat{T} + \hat{V}_{ee} + \hat{v}_{ext} \middle| \Psi_0 \right\rangle$$

(13)

whence the DFT correlation energy E_c is defined as:

$$E_c = \left\langle \Psi_0 \middle| \hat{T} + \hat{V}_{ee} \middle| \Psi_0 \right\rangle - \left\langle \Phi_s \middle| \hat{T} + \hat{V}_{ee} \middle| \Phi_s \right\rangle$$
$$= \left\langle \Psi_0 \middle| \hat{T} + \hat{V}_{ee} + \hat{v}_{ext} \middle| \Psi_0 \right\rangle - \left\langle \Phi_s \middle| \hat{T} + \hat{V}_{ee} + \hat{v}_{ext} \middle| \Phi_s \right\rangle$$

(14)

The second equality follows because Ψ_0 and Φ_s correspond to the same electron density, see Eq. (8). Replacement of the KS determinant by the HF determinant in the second part of Eq. (14) would recover the conventional quantum chemistry correlation energy introduced earlier. While similar in form, the two correlation energies differ because the model wave functions differ. The HF determinant, by definition, is the determinant yielding the lowest total energy expectation value with respect to the original molecular Hamiltonian. Therefore the variational principal says that the KS determinant must lead to a higher (or at least equal) expectation value and thus the DF correlation energy must be equal or larger in magnitude (usually larger) than the traditional quantum chemical value for the same molecule.

Each electron density ρ defines a set of energies T_s, U, E_x, and E_c. The energies thus are functionals $T_s[\rho]$, $U[\rho]$, $E_x[\rho]$, and $E_c[\rho]$ of ρ. For U this is obvious; the explicit functional dependence on ρ is known [Eq. (6)]. For the other energies the case of a v-representable electron density is considered. Then, according to the first HK theorem, both an interacting and a non-interacting Hamiltonian are uniquely determined by the density via the corresponding external and effective potentials, respectively. The ground states corresponding to these Hamiltonians then in turn also are determined by the density ρ and thus can be considered as functionals $\Psi[\rho]$ and $\Phi[\rho]$ of ρ. The wave functions $\Psi[\rho]$ and $\Phi[\rho]$ now yield $T_s[\rho]$, $E_x[\rho]$, and $E_c[\rho]$ as expectation values of the respective operators. We note that functionals $T_s[\rho]$, $E_x[\rho]$, and $E_c[\rho]$ can be defined for any electron density, not just v-representable ones, in the more general constrained-search formulation of DFT [46,47] which will not be discussed here.

The ground-state energy E_0 of a molecule is given as the sum of the functionals $T_s[\rho]$, $U[\rho]$, $E_x[\rho]$, and $E_c[\rho]$ evaluated at the ground-state density ρ_0, plus the interaction energy with the external potential $\int d\vec{r} v_{ext}(\vec{r}) \rho_0(\vec{r})$. (Just add the definitions of the energies and note that $\Psi[\rho_0] = \Psi_0$ and $\Phi[\rho_0] = \Phi_s$.) If the effective potential v_s [Eq. (3)] for the molecule of interest were known, then the KS orbitals would make all contributions to E_0 directly calculable except for the correlation energy E_c. The exact form of E_c is unknown, so with the equations displayed thus far, one still would have to calculate the full ground state Ψ_0, exactly the task to be avoided. In fact, good approximations are known for the density functional $E_c[\rho]$ (see Sect. 3). Though neither $T_s[\rho]$ nor $E_x[\rho]$ are known as explicit functionals of ρ, both can be calculated directly from the KS orbitals, hence approximate functionals are not required. (For reasons given later, in practice $E_x[\rho]$ often is evaluated via approximate density functionals and not exactly via the KS orbitals.) $U[\rho]$, the Coulomb repulsion, of course, can be calculated exactly from Eq. (6).

We pause to recapitulate what has been achieved. The HK variational density functional promised by the second HK theorem has been recast into contributions all of which are calculable from KS quantities:

$$E[\rho] = T_s[\rho] + U[\rho] + E_x[\rho] + E_c[\rho] + \int d\vec{r} v_{ext}(\vec{r})\rho(\vec{r}) \tag{15}$$

The minimum value of this functional, at ρ_0, is the exact ground-state energy E_0. Put into words, the neat thing about Eq. (15) is that it has an exact value at the exact minimum, yet can be manipulated entirely in terms of KS model quantities.

2.1.3
Effective Potentials

The major task remaining is to calculate the KS orbitals, which requires knowledge of the effective potential v_s in the single-particle KS equation [Eq. (4)]. Let
us return for a moment to the HF approximation. The effective Hamiltonian for
the HF orbitals follows from the requirement that the molecular energy, a scalar,
be stationary with respect to variations of the HF determinant. This condition
leads directly to the HF equations. Evidently the KS formalism is different because v_s must reproduce the ground-state density, a function of three-dimensional space, via the KS determinant which v_s generates. Formal justification of
the procedure to be described is readily available in the literature [26,28]. As just
suggested, motivation is provided by the variational nature of Eq. (15). Thus the
idea of the KS procedure is to vary $E[\rho]$ over densities ρ for those terms where
that dependence is explicit and over the orbitals in the associated determinant
in those terms where that is more convenient. If Eq. (15) were available in its exact form, variation would give an effective Hamiltonian which corresponds to
the exact ground-state energy and density, so the potentials in that Hamiltonian
are the ones sought.

Variation with respect to the density brings into play the functional derivatives
of the terms in $E[\rho]$. A functional derivative gives the first-order change of a functional due to a change in the function which is its independent variable, here a
change in the function ρ. For the familiar Coulomb repulsion energy U we have:

$$\delta U = \int d\vec{r}\, \frac{\delta U[\rho]}{\delta \rho(\vec{r})} \delta\rho(\vec{r}) = \int d\vec{r} \int d\vec{r}' \frac{\rho(\vec{r}')}{|\vec{r}-\vec{r}'|} \delta\rho(\vec{r}). \tag{16}$$

which gives the classical Coulomb potential u for charge density ρ as the functional derivative of $U[\rho]$:

$$u([\rho];\vec{r}) = \frac{\delta U[\rho]}{\delta\rho(\vec{r})} = \int d\vec{r}' \frac{\rho(\vec{r}')}{|\vec{r}-\vec{r}'|}. \tag{17}$$

Similarly, the exchange potential $v_x([\rho];\vec{r})$ and the correlation potential
$v_c([\rho];\vec{r})$ are:

$$v_x([\rho];\vec{r}) = \frac{\delta E_x[\rho]}{\delta\rho(\vec{r})} \tag{18}$$

$$v_c([\rho];\vec{r}) = \frac{\delta E_c[\rho]}{\delta\rho(\vec{r})} \tag{19}$$

The effective potential v_s is the sum of the external potential plus the three
potentials just listed and the single-particle KS equation, Eq. (4), has the
form:

$$\left[\left(-1/2\right)\vec{\nabla}^2 + v_{ext}\left(\vec{r}\right) + u\left(\left[\rho\right];\vec{r}\right) + v_x\left(\left[\rho\right];\vec{r}\right) + v_c\left(\left[\rho\right];\vec{r}\right)\right]\varphi_i\left(\vec{r}\right) = \varepsilon_i\varphi_i\left(\vec{r}\right) \qquad (20)$$

which is the equation actually solved by KS methods. Note that, like the HF equations, the KS equation must be solved self-consistently because the potentials depend on the electron density ρ, hence on the KS orbitals.

The N-electron KS equation which is equivalent to Eq. (20) is:

$$\left[\hat{T} + \hat{v}_{ext} + \hat{u}\left[\rho\right] + \hat{v}_x\left[\rho\right] + \hat{v}_c\left[\rho\right]\right]\Phi_s = E_s\Phi_s \qquad (21)$$

Comparison with Eq. (11) shows that the electron–electron interaction operator in the original full Hamiltonian $\hat{H}$ is replaced by the sum of $u([\rho];\vec{r})$, $v_x([\rho];\vec{r})$, and $v_c([\rho];\vec{r})$. Thus the effects on ρ of the complicated interactions of the electrons in the molecule are reproduced by superficially simple correction terms to the external potential. These corrections are simple only in the sense of being ordinary single-particle potentials. Except for $u([\rho];\vec{r})$ they are, however, not at all simple in their functional dependence on ρ. Other than for abstract systems (primarily the uniform electron gas), the explicit functional dependence is largely unknown and therefore approximations must be constructed.

For the correlation potential the need for approximation is no great surprise, because the density dependence of the correlation energy is not known explicitly. Highly accuracte, flexible treatment of the correlation problem is still a major focus of modern quantum chemistry. Even relatively straightforward treatments involve quite complicated approximate wave functions which do not give the explicit density dependence of E_c. Without that dependence, the functional derivative cannot be executed. The most important properties and approximations for the correlation functional E_c are discussed in Sect. 3. Approximations for v_c follow as functional derivatives of whatever approximation is used for E_c.

All other parts of the energy either are known explicitly in terms of the density (U and E_{ext}) or can be calculated via the KS orbitals (T_s and E_x). However, the explicit density dependence of the exchange functional $E_x[\rho]$ is also unknown (except for some highly idealized systems [48–50]). The known explicit dependence is in terms of the KS orbitals [Eq. (9)]. This, however, is not enough to make the exact v_x accessible simply. For that one would need to carry out the functional derivative form of the chain rule, hence would need the explicit functional derivative of the KS orbitals with respect to ρ, which also is generally unavailable. The common route out of this dilemma is to approximate E_x, usually together with E_c to form E_{xc}, and gain some useful error cancellation. This strategy also has the benefit of reducing the complexity of the integrals (matrix elements) which must be computed (whence arises the statement that the normal KS procedure formally scales as N^3 vs. N^4 for HF). It also assures a consistent variational treatment of exchange, whereas if E_x were treated exactly and v_x approximately, the potential would not be the exact functional derivative of the energy, i.e. Eq. (18) would be violated. That inconsistency would cause problems for calculating energy gradients with respect to the nuclear coordinates (because the Hellmann-Feynman theorem no longer would hold).

An alternative route which gets the exact KS v_x (not to be confused with the non-local HF exchange potential) from the orbital representation of E_x has so far been realized only for atoms [51] and, in one case, for semiconductors [52,53]; for molecules, approximate numerical procedures in this vein have been studied [54,55].

Oddly, but importantly, there is a problem hidden in the seemingly innocuous form of U that is corrected by using the exact E_x and v_x but is not well-treated by most approximate forms of E_x and associated v_x. Called spurious self-interaction, it is easy to understand. Think about a one-electron system, for example, the H atom, and evaluate Eq. (6), which defines U. The result is a completely unphysical non-zero Coulomb energy, because the electron interacts with itself. For the exact E_x, there is a compensating self-exchange which is also unphysical but exactly cancels the self-repulsion. The consequences of improper cancellation of self-repulsion by approximate forms of E_x are many. In anticipation of later remarks, these effects include improper results for anions, artificially high atomic total energies (and therefore, artificially overbound molecules), and shifted one-electron levels. One example is the reduction of the gap between highest occupied and lowest unoccupied MO eigenenergies ("HOMO-LUMO" gap), a point to which we return later.

Let us summarize. A KS calculation for the ground state consists of two steps (after choosing an approximation for E_x and for E_c): First the KS equation is solved self-consistently for the MOs and eigenvalues. Second the ground-state density is constructed according to Eq. (7) and the ground-state energy is calculated with Eq. (15). Obviously any other properties of the ground state, e.g. the dipole moment, which depend exclusively on the density may be calculated. It is commonplace, though not rigorous, to calculate other properties which are not simply functions of the density alone as expectation values with respect to the KS determinant. (The observant reader will note that we have said almost nothing about excitation energies yet.)

From the present point of view, typical contemporary KS calculations involve three approximations (beyond such things as basis set truncation). Both the exchange and correlation potentials are approximated, hence there are inaccuracies in the MOs. Those errors propagate to all parts of the calculated ground-state energy, through the orbitals themselves and via the density ρ. Second, E_x is not calculated from the exact orbital-dependent expression Eq. (9) but from the approximate $E_x[\rho]$ whose functional derivative is the v_x being used. Thirdly, the accuracy of E_c is also limited by use of an approximate density functional.

2.1.4
Kohn-Sham Eigenvalues

We turn to the interpretability of the KS MOs and associated eigenvalues, a topic about which much has been written [26,56–59]. Our emphasis is on some of the salient aspects from the viewpoint of applying DF methods through current DF codes.

Recall, for context, that HF eigenvalues are interpreted by recourse to Koopman's theorem [25] which justifies the approximation of ionization potentials (i.e. the difference between total energies of the ground state and an ionic state) by the HF eigenvalue of the MO from which the electron is removed. Analogously, excitation energies may be approximated as differences of HF eigenvalues.

KS eigenvalues do not afford such a simple interpretation, except for the highest occupied MO; with the exact E_c and E_x that eigenvalue equals the electron removal energy or (first) ionization energy. KS eigenvalue differences are well-defined approximations of excitation energies [60] in the context of what is called the adiabatic connection formalism [61–63] (see Sect. 3.1). There are numerical studies of the eigenvalues of near-exact KS effective potentials which show that KS eigenvalue differences are surprisingly good estimates of excitation energies that can easily be improved even further by a simple perturbation theory approach [60,64,65]. As an aside about the current state of research in DFT, those near-exact potentials are derived from very-high-quality calculated molecular densities.

The KS eigenvalues also obey the Janak-Slater theorem [56,57], which comes from a fairly simple but in some ways subtle idea. The concept is to generalize the KS scheme presented above to allow for continuously variable occupation numbers n_i (between zero and two) of the KS orbitals in the density: $\rho = \sum_i n_i \varphi^*_i \varphi_i$. The subtlety of course is that the KS ground state no longer is a single determinant. Leaving that conceptual issue aside, the Janak-Slater theorem states that the KS eigenvalues are the derivative of the total energy with respect to the corresponding occupation number:

$$\varepsilon_i = \frac{\delta E[\rho]}{\delta n_i} \qquad (22)$$

This expression gives the differential cost in energy for removing electrons from or adding them to a KS MO. In combination, the Janak-Slater theorem and the fact that the KS MOs build the density provide a powerful semi-quantitative tool to connect with successful interpretive uses of MOs. Chemical concepts long have been understood and interpreted in terms of MOs, a natural extension of the picture of properties across the periodic table which emerges from little more than the Pauli principle in the central field atom approximation. For many interpretive purposes, the most useful MOs turn out not to be the rigorous HF or post-HF ones [24,25] but the semi-empirical orbitals from Hückel [66] theory, extended Hückel theory [2,67], and related constructs. A key interpretive feature is that electrons redistribute from higher to lower energy extended Hückel orbitals that form when two moieties interact. The Janak-Slater theorem justifies similar redistribution in a semi-quantitative analysis of KS eigenvalues. Furthermore, the KS MOs are direct ingredients in the construction of a physical observable, the electron density. Since structure determination (e.g. by X-rays) is a determination of electron density, electron redistribution in KS theory is directly and intuitively connected with molecular structure.

Qualitative and semi-quantitative characterization of electron redistribution in terms of electronegativity, hardness, and softness has been known in the literature since the work of Mulliken [26,68] and Pearson [26,69,70]. Much more recently these have been put on a quantitative basis within DFT. The route is by a seemingly arcane exercise. In any real molecule (at least in the dilute gas phase) the total number of electrons N is a fixed integer. Nevertheless, a variety of motivations outside the scope of this article led to the extension of DFT to continuously variable N. One obvious aspect of such extensions is that they would be concerned with reactivity, since a change in electron count often affects the reactivity of a system. We give only the briefest of sketches: detailed reports are available elsewhere [34,71–73].

A bit of terminology is inescapable. To use the total electron number as a variable but end up with a specified number of electrons in the molecule requires a constraint and therefore a Lagrange multiplier, call it Λ. This function plays the role of a chemical potential for the theory, just as a real chemical potential keeps the right number of particles in a chemical system. For the exact effective potential v_s, Parr and coworkers [71] introduced the equation

$$\Lambda = \frac{\delta E}{\delta N}\bigg|_{v_s} \approx -\frac{I+A}{2} = -\chi \tag{23}$$

where E is $E[\rho]$ in Eq. (15), χ is the Mulliken electronegativity, and I and A are the first ionization potential and electron affinity, respectively. The step for the approximation is a subtle and bold one: the derivative is discontinuous at integer numbers of electrons and the approximation replaces that discontinuity with an average of the two values at either side of the discontinuity. Contemporary approximate functionals (see Sect. 3.2) in fact have continuous derivatives as well. Also for the exact v_s we have already noted that

$$\varepsilon_{HOMO} = -I \tag{24}$$

Here the HOMO eigenvalue is for the ground state. The global hardness is

$$\eta = \frac{1}{2}\frac{\partial^2 E}{\partial N^2}\bigg|_{v_s} \approx \frac{I-A}{2} \tag{25}$$

and the approximation sign arises from averaging discontinuous derivatives. Together these two indices provide the beginnings for a quantitative description of electron flow in the formation of bonds. (There is also a considerable variety of position-dependent or "local" indices which we will not discuss; see [26].) Consider the schematic reaction A+B→AB and suppose $\Lambda_A > \Lambda_B$. Then other things being equal and the structure of A and B being discernible after the reaction (as is commonly the case), the electron flow during the reaction should be from A to B. By expanding in powers of the incremental electron transfer ΔN, one finds (to first order) the quantitative estimate

$$\Delta N \approx \frac{\Lambda_A - \Lambda_B}{2(\eta_A + \eta_B)} \tag{26}$$

Arguments have been put forward for the superiority of these indices over reasoning based on Hückel theory, either simple or extended [71].

A useful computational trick was devised by Slater [56,74] to get values for I and A when using an approximate v_s, which is to say, most of the time. For such potentials Eq. (24) typically is not accurate enough to be useful in Eqs. (23) or (25); recall, for example, the comments above about the consequences of inadequate self-repulsion cancellation. The trick is to calculate the HOMO and LUMO eigenvalues for a system with half an electron removed or added (relative to the ground-state occupancy), respectively:

$$I \approx -\varepsilon_{HOMO}(n_{HOMO} = 1/2)$$
$$A \approx -\varepsilon_{LUMO}(n_{LUMO} = 1/2) \tag{27}$$

Although this recipe has nothing to do with chemical transition states, it is called "the Slater transition state" in the literature [56,74].

2.2
Extension to Open-Shell Systems

2.2.1
Spin Density Functional Theory

To here it was implied that the KS potential v_s and the KS equation always exhibit the full molecular symmetry with respect to spin. The point of the remark is that, neglecting spin-orbit interaction for the moment, the full molecular Hamiltonian is always fully spherically symmetrical in spin space and the KS equation developed above exhibits no explicit spin dependence. Two spin orbitals, one for spin-up, the other for spin-down, result from one spatial orbital. In DFT this is called the "non-spin-polarized" treatment; the HF analogue is the restricted HF (RHF) formalism.

Clearly the corresponding spin-up and spin-down densities are not given correctly by the non-spin-polarized KS determinant for all systems. The point is obvious for open-shell systems. There is no constraint on the non-spin-polarized KS determinant to force it to yield the true spin components. This might not be such a serious shortcoming if it were not accompanied by relatively poor results for the energetics of open-shell systems. Therefore one resorts to spin density functional theory. Note, the failure of the non-spin-polarized treatment to yield good energies is not from fundamentals but from shortcomings of the non-spin-polarized exchange and correlation functionals. In fact, it is possible to formulate ordinary DFT with constraints to give the correct values of S^2 and S_z as well as the energy [75] but this form of the theory is not used in practice so far as we are aware.

The essential added feature in the KS formalism for spin density functional theory is that the spin-polarized KS wave function is required to yield spin-up and spin-down electron densities which individually are identical to those of the real molecular ground state. The result is a pair of coupled single-particle KS equations for spin-up and spin-down orbitals. Both v_x and v_c are spin-labeled, as are the orbitals and eigenvalues. Neither the two equations nor the spatial parts of the orbitals are identical. This complication both increases computational effort and prevents a proper symmetry classification of the KS wave function with respect to the spin quantum numbers: the KS wave function is said to be spin-contaminated. This is a serious disadvantage of spin density functional theory which is offset by substantially more accurate energetics. Procedurally, spin density functional theory, frequently called the "spin-polarized" treatment, bears much formal similarity to the unrestricted HF (UHF) formalism and shares some of its problems.

Otherwise, the discussion already given about the characteristics of KS theory in practice applies fairly generally to spin-polarized treatments, with one additional caveat. Because there are two coupled equations, their actual self-consistent solution in a typical computer code can be seriously, sometimes frustratingly slow. The problem is iteration-to-iteration oscillation of charge between nearly degenerate spin-up and spin-down KS MOs. There are many numerical techniques for dealing with such problems and much art as well, but no sure-fire recipes.

Multiplet splittings are not accessible, in general, within present-day formulations of DFT, since excited states (see discussion, end of Sect. 2.2.2) are involved. A reasonably effective approximate strategy has been suggested which, in its simplest incarnation, combines a spin-polarized treatment with determinants constructed from KS orbitals to evaluate the singlet-triplet splitting at reasonable accuracy [76–78]. Recently, this procedure has been applied successfully to the calculation of multiplet splittings of TM complexes [79].

2.2.2
Spatial Symmetry Aspects

Open-shell molecules bring in not only spin-polarization but other problems related to spatial symmetries and degeneracies. We give a very brief discussion oriented to the current practice with commonly available computer codes.

In open-shell systems, the KS approach suffers from the following problem: an open-shell electron density in general does not exhibit the full symmetry of the molecular Hamiltonian (unless the spatial parts of the open-shell MOs belong to one-dimensional irreducible representations). For simplicity, consider the ground state of atomic Cl, ^2P; many other examples may be constructed for TM complexes. The associated ground-state density does not have the full spherical symmetry of the atomic Hamiltonian but cylindrical symmetry. The effective potential of the KS Hamiltonian derived for a given ground state has in general only the symmetry of the electron density. Therefore the KS determinant from the KS Hamiltonian can only be classified according to this lower symmetry and not, as would be desirable for interpretative purposes, according to the full molecular symmetry.

For atoms and certain other systems the problem can be side-stepped by classifying the system states within the lower symmetry group right from the start. For the example of Cl, this would mean using $D_{\infty h}$ symmetry instead of full spherical symmetry. The state ^2P then splits into $^2\Sigma_u^-$ and $^2\Pi$; the density associated with the former state is totally symmetric with respect to $D_{\infty h}$. With the exception of atoms, this scheme usually is not used as the KS equation then has a lower symmetry than the physical system. For a more general but quite formal approach along these lines see Ref. [80].

The common practical way to eliminate the conceptual problem is to replace the electron densities for the degenerate states by their average. The averaged electron density always exhibits the full molecular symmetry, hence the problem is alleviated. For atoms this is the well-known central field approximation. The procedure has the advantage that the full system symmetry can be exploited in the determination of the KS orbitals. Issues of justification and consequences then arise. A formal justification can be given by a symmetrized density functional theory in which the basic quantity is the totally symmetric part of the density; the latter quantity is given simply by the average electron density just described [81].

Now comes the so-called symmetry dilemma [82]. The most prominent example is probably Cr_2, but the problem already is evident in the homolytic dissociation of H_2. The non-spin-polarized KS value for the H_2 dissociation energy is much too high, despite the fact that H_2 is a closed-shell singlet for any bond distance. Spin-polarized KS calculations yield a reasonable potential energy curve for the dissociation. However, the resulting spin density breaks the molecular symmetry: predominantly spin-up on one center and the reverse on the other. This result contradicts both the fact that the spin density of a $^1\Sigma_g^+$ state is totally symmetric and the claim that spin density functional theory gives the correct spin density. It appears necessary to accept either a qualitatively correct spin density and a bad dissociation energy or a reasonable dissociation energy and a qualitatively wrong spin density. This dilemma occurs in the many situations in which one wants to describe a biradicaloid system or a molecule of analogous electronic structure with current KS methodology.

The original version of this problem originally occurs in HF theory and was so named by Löwdin [83]. It is almost the same in KS theory so we only sketch it. The difficulty is that in the limit of large bond lengths, the singlet KS orbitals form the molecular equivalent of two non-spin-polarized atoms with half an electron each of spin-up and spin-down density localized on each atom. As a result there is a spurious Coulomb repulsion that ruins the dissociation energy. In an exact KS method this unphysical Coulomb interaction would be canceled by the correlation energy. The available approximate correlation functionals, however, are not able to do this (nor is it easy to see how to build functionals that would). Thus the symmetry dilemma is not an essential problem of DF theory as such, but can be traced back to shortcomings of the present approximations for the correlation density functional.

For H_2, both spin-polarized and non-spin-polarized calculations for near-equilibrium bond distances yield essentially the same KS orbitals. Beyond a certain distance, inversion symmetry is broken for the spin-polarized KS orbitals and they start localizing on one or the other H nucleus [63,84]. In the limit of large distances there is a spin-up electron localized about one nucleus and one with spin-down on the other. The resulting KS determinant yields the energy of two separated hydrogen atoms. (We note that, in contrast to the spin-polarized KS method, the unrestricted HF method formally is not required to yield the correct spin-density.)

It is instructive to consider the qualitative features of the true wave function for dissociated H_2 in comparison with the KS wave function. For infinite separation of the two H atoms, the true wave function of the $^1\Sigma_g^+$ ground state can be written as a linear combination of two Slater determinants or in valence-bond form:

$$\Psi_0 = \frac{1}{\sqrt{2}}\left[\left|\sigma_g\bar{\sigma}_g\right| - \left|\sigma_u\bar{\sigma}_u\right|\right] = \frac{a(1)b(2)+b(1)a(2)}{\sqrt{2}}\frac{\alpha(1)\beta(2)-\beta(1)\alpha(2)}{\sqrt{2}} \quad (28)$$

Here, $\sigma_{g/u}=(a\pm b)/\sqrt{2}$ are the symmetrized linear combinations of the ground-state orbitals of each H atom, a and b; as usual, a bar indicates an orbital with spin-down direction, i.e. $\sigma_g\equiv\sigma_g\alpha$ and $\bar{\sigma}_u\equiv\sigma_u\beta$. In a minimal basis, we have four H 1s spin orbitals (spin-up and -down on each atom). The first determinant is essentially the KS ground state. For distances near the equilibrium of H_2 this contribution dominates the true wave function (albeit with distorted atomic functions) but for large distances that is not the case and we have a true multi-determinantal wave function. Thus we may attribute the problem of describing the dissociated H_2 correctly via the KS procedure to the fact that the true wave function cannot be well represented by a single Slater determinant. This finding can be generalized. States that inherently comprise several Slater determinants often cause problems in KS calculations. In the present example, this situation is connected to the symmetry, but occurs more commonly for TM compounds due to near-degeneracies among various electron configurations. We are back to the static correlation problem discussed in the Introduction (Sect. 1.1).

A final aspect of spatial symmetry is the treatment of excited states. It is important in principle (less so in practice) that the KS method can be extended to the lowest energy molecular excited state of each different symmetry because the HK theorems can be generalized to cover this situation. One might hope, therefore, to determine the resulting symmetry-dependent KS orbitals in the usual way, but things are not so simple. First, the exchange and correlation functionals become symmetry dependent [63]. As far as we are aware, that dependence is neglected in practice, thereby significantly limiting the applicability of the formalism. Second, although frequently invoked as a justification for the treatment of the energetically lowest state of each symmetry, the problem of reduced symmetry which we noted above for open shells also shows up for all excited states without the full symmetry of the Hamiltonian. Now this problem is a severe conceptual one: The generalization of the HK theorem for the lowest states of each symmetry requires a classification of the KS wave function according to symmetry which, however, in general is not possible. Note that this class of difficulties therefore applies particularly if the ground state does have the full symmetry of the molecular Hamiltonian, since all the excited states which can be treated must be orthogonal by symmetry. Again, the common method to overcome the problem is to average over the electron densities of all degenerate states, i.e. to employ a symmetrized formalism.

The spin-polarized formalism also can be used for the lowest excited states of symmetries different from the ground state. Averaging over the symmetry partners of an energy level is carried out only with respect to symmetries in ordinary space, not in spin space. Thus, only states with the same number of spin-up and spin-down electrons are averaged. As before, the symmetry dependence of the spin density functionals is neglected in practice.

3
Exchange and Correlation Holes; Functionals

3.1
Physical Constraints on Functionals from Properties of Electron Holes

Successful application of the KS procedure evidently depends on having good approximations to the correlation functional $E_c[\rho]$. Good approximations for $E_x[\rho]$, though not essential formally, are extremely important computationally for consistency between E_x and v_x; recall earlier discussion. (Historically it was also true that matrix elements of $v_x[\rho]$ were easier to compute than those of the HF-like exchange operator which derives from the explicitly orbital-dependent KS E_x.) Neither the HK theorems nor the KS procedure give prescriptions for determining such approximations. They are built therefore by chemical and physical reasoning, insight from "toy" problems, and by constraining them as much as possible to obey the known formal properties of E_c and E_x.

Three classes of constraint are important, broadly speaking: general properties of the exchange-correlation hole, its coordinate scaling conditions, and asymptotics. In a sense the latter are aspects of the first. Hole constraints also are perhaps the most physical or interpretable.

Because the full Hamiltonian $\hat{H}$ includes only one- and two-electron operators, it long has been understood [36] that the expectation value of $\hat{H}$ with respect to the exact Ψ_0, Eq. (13), reduces to integrals over two closely related distribution functions. We actually need only one quantity, the diagonal part $\rho_2^{\sigma_1\sigma_2}(\vec{r}_1,\vec{r}_2)$ of the second-order density matrix. This quantity gives the probability of finding an electron with spin σ_1 at the position $\vec{r}_1$ and another electron with spin σ_2 at position $\vec{r}_2$. Denoting all the other coordinates and spins collectively as $\mathbf{R}$ and $\mathbf{s}$, we have

$$\rho_2^{\sigma_1\sigma_2}(\vec{r}_1,\vec{r}_2) = N(N-1)\int d\mathbf{R}d\mathbf{s}\Psi^*(\vec{r}_1\sigma_1\vec{r}_2\sigma_2\mathbf{Rs})\Psi(\vec{r}_1\sigma_1\vec{r}_2\sigma_2\mathbf{Rs}) \tag{29}$$

(The normalization of $\rho_2^{\sigma_1\sigma_2}(\vec{r}_1,\vec{r}_2)$ differs by a factor of 2 from the density matrix literature.) Because it involves only two electrons at a time, the same reduction occurs in the exact total electron-electron Coulomb energy:

$$E_{ee} = \left\langle \Psi_0 \middle| \hat{V}_{ee} \middle| \Psi_0 \right\rangle = \frac{1}{2} \int d\vec{r}_1 d\sigma_1 \int d\vec{r}_2 d\sigma_2 \rho_2^{\sigma_1\sigma_2}(\vec{r}_1,\vec{r}_2) \frac{1}{|\vec{r}_1 - \vec{r}_2|} \tag{30}$$

Clearly the spin "integrals" are simple sums and an important object is the pair density

$$\rho_2(\vec{r}_1,\vec{r}_2) = \sum_{\sigma_1\sigma_2} \rho_2^{\sigma_1\sigma_2}(\vec{r}_1,\vec{r}_2) \tag{31}$$

so

$$E_{ee} = \frac{1}{2} \int d\vec{r}_1 d\vec{r}_2 \rho_2(\vec{r}_1,\vec{r}_2) \frac{1}{|\vec{r}_1 - \vec{r}_2|} \tag{32}$$

Because the Coulomb interaction [integrand of E_{ee}, Eq. (32)] depends only on the separation of the two electrons, not on their absolute position in the system nor angular orientation, a simple variable change is enough to show that only the volume and angle average of ρ_2,

$$\rho_2(r) = \frac{1}{4\pi} \int d\Omega \int d\vec{r}_1 \rho_2(\vec{r}_1,\vec{r}_2 + \vec{r}), \tag{33}$$

is needed, so that at least some of the details of the pair density need not be reproduced to have a good approximation.

To get constraints on E_x and E_c, we split off pieces in ρ_2 to match the decomposition of the explicitly coulombic parts of $U + E_x + E_c$:

$$\rho_2(\vec{r}_1,\vec{r}_2) \equiv \rho(\vec{r}_1)\left[\rho(\vec{r}_2) + \rho_x(\vec{r}_1,\vec{r}_2) + \rho_c(\vec{r}_1,\vec{r}_2)\right] \equiv \rho(\vec{r}_1)\left[\rho(\vec{r}_2) + \rho_{xc}(\vec{r}_1,\vec{r}_2)\right] \tag{34}$$

Obviously there are corresponding spin-labeled versions. The factorization on the right-hand side of Eq. (34) has a simple conditional probability interpretation: given an electron "at" $\vec{r}_1$ (i.e. in $d\vec{r}_1$ surrounding $\vec{r}_1$), the sum in brackets is proportional to the probability that there is also an electron at $\vec{r}_2$. The first term in Eq. (34) gives U, which is of purely classical electrostatic form; cf. Eq. (6). If it were the only term, then the conditional probability of an electron at $\vec{r}_1$ given another at $\vec{r}_2$ would simply be the product of the two electron densities. This cannot be right because if we know where one electron is, then there are only $N-1$ others to account for. Thus the first product in Eq. (34) double counts one electron; this is the spurious self-repulsion in U already discussed. It is corrected by $\rho_x(\vec{r}_1, \vec{r}_2)$ the exchange hole (density), which both corrects the double counting and prevents two electrons with parallel spin from having the same position (Pauli exclusion). The third contribution, ρ_c in Eq. (34), is the correlation hole;

see remarks below. The two quantities are "holes" because they reduce the probability below $\rho(\vec{r}_2)$ in the vicinity of $\vec{r}_1$; pictorially they put a hole in the density. The sum, $\rho_{xc}=\rho_x+\rho_c$, is often referred to as the exchange-correlation hole.

The exact form of $\rho_x(\vec{r}_1, \vec{r}_2)$ in terms of the KS orbitals is known from Eq. (9). Anti-parallel spins are unaffected by exchange, so their exchange hole is trivially zero everywhere:

$$\rho_x^{\alpha\beta}(\vec{r}_1,\vec{r}_2)=0 \tag{35}$$

The hole for parallel spins integrates to one electron and since exchange keeps one parallel spin electron away from another, ρ_x for them is never positive:

$$\rho_x^{\sigma\sigma}(\vec{r}_1,\vec{r}_2)\leq 0$$
$$\int d\vec{r}_2 \rho_x^{\sigma\sigma}(\vec{r}_1,\vec{r}_2)=-1 \tag{36}$$

The second property in Eq. (36) is the one which assures that any proper approximation to E_x will cancel the self-repulsion. Recall (Section 1.1) that dynamical correlation, on the other hand, moves electrons of both spins away from any given electron in order to reduce the cost of Coulomb repulsion as much as possible. Thus ρ_c must rearrange probability but, unlike ρ_x, it does not reduce the count of interacting electrons. ρ_c also must account for static correlation, the multi-reference effects which arise from near degeneracies (again, recall Section 1.1). Therefore ρ_c is a function which for electron "1" at fixed position integrates to zero over the coordinates of electron "2" for both parallel and anti-parallel spin cases:

$$\int d\vec{r}_2 \rho_c^{\sigma_1\sigma_2}(\vec{r}_1,\vec{r}_2)=0 \tag{37}$$

We give a few brief remarks about the other two kinds of constraints. A failure in the asymptotic behavior of the current approximate correlation functionals has already been mentioned: the symmetry dilemma (even for H_2). Another problem related to asymptotics is the form of the large distance (large-r) tail of v_x+v_c. Extensive analysis has been given by many authors but the basic constraint is simple: the total potential for an electron removed far from an atom must go as $1/r$. Examples of overt attention to this behavior in an approximation for E_x and v_x can be found in Refs. [85,86]. We will consider the asymptotic behavior of specific approximations below. Coordinate scaling is the formalization of the dimensional analysis which shows, for example, that E_x must be of the form of an integral of $\rho^{4/3}$ times a dimensionless function of ρ. Levy has been the leading exponent of coordinate scaling; a long list of constraints is found in Ref. [47] along with references to the original work. Many properties for the two kinds of electron holes and their sum are known. Reviews and references to the original literature are given in Refs. [47,87].

Before proceeding, it is useful to introduce one more tool that has played a major role in understanding exchange and correlation in DFT, the adiabatic connection expression for E_{xc}. Early on we introduced the exact wave function Ψ_0 and the KS non-interacting wave function Φ_s that correspond to the same electron density. A comparison of the exact N-particle Hamiltonian $\hat{H}$, Eq. (11), with the N-electron KS equation, Eq. (21), suggests that we can get from one to the other by turning off the Coulomb interaction V_{ee} smoothly while turning on the effective potential so as to keep the density ρ fixed. With this idea in mind, define

$$W_{xc}[\rho;\lambda] = \left\langle \Psi^{\rho}_{\min}(\lambda) \middle| \hat{V}_{ee} \middle| \Psi^{\rho}_{\min}(\lambda) \right\rangle - U[\rho] \tag{38}$$

where the peculiarly labeled wave function is the proper N-particle wave function which minimizes the expectation value of the modified Hamiltonian $\hat{T} + \lambda\hat{V}_{ee}$ and gives the specified, fixed ρ at the value of the coupling constant λ ($0 \leq \lambda \leq 1$). Then it can be shown that

$$E_{xc}[\rho] = \int_0^1 d\lambda\, W_{xc}[\rho;\lambda] \tag{39}$$

Thus, the exchange-correlation functional, which includes the kinetic energy contribution $\left\langle \Psi_0 \middle| \hat{T} \middle| \Psi_0 \right\rangle - T_s[\rho]$, comes from "summing" potential energy contributions W_{xc} of the electron–electron interaction ranging from non-interacting to full physical strength. Obviously Eq. (39) could be used to build approximate exchange-correlation functionals if it turned out to be easier or more productive to approximate the integral than to build E_{xc} directly. One line of argument which suggests at least the form of the recently popular "hybrid" approximate functionals makes use of the adiabatic connection formula; see Sect. 3.2.2.

3.2
Approximate Functionals

To have a usable KS equation we must have v_x and v_c while the energetics require E_x and E_c. Today there is vigorous research to construct and study the potentials v_x and v_c which, when used in the KS equation, produce the densities corresponding to extremely high quality wave functions. Several such potentials have been proposed [88]. Nevertheless, the reality for utilization of DFT in chemistry today is that E_x and E_c are approximated and their functional derivatives taken to give the corresponding v_x and v_c. In view of what may seem to be a bewilderingly large number of approximate E_x and E_c, a sweeping classification may be helpful. First, usage today splits into two large classes with respect to E_x: approximations which use only an E_x functional that explicitly depends on the density (or spin densities) and those which also include a contribution from the explicit orbital-dependent form. Approximations in the latter category are called "hybrids" while those in the former category commonly are called "local and semi-local". The "semi-local" part is misleading, as we shall soon see, so we will use a more literal term: "local and gradient corrected". Second, there are both first-principles functionals and functionals parameterized to some data source in both categories.

3.2.1
Local and Gradient-Corrected Approximate Functionals

A general point to notice in this category is that exchange and correlation are implemented together (although modern exchange and correlation contributions frequently are designed separately in order to satisfy the constraints already mentioned).

Members of the oldest branch of this category are the closely related variants of the Local (Spin) Density Approximation, LDA and LSDA. The general LSDA is

$$E_{xc}^{LSDA} = \int d\vec{r} \rho(\vec{r}) \mu_{xc}[\rho^{\alpha}(\vec{r}), \rho^{\beta}(\vec{r})] \tag{40}$$

Here, ρ^{α}, ρ^{β} are the spin-up and spin-down densities, defined analogously with Eq.(7), except that the summation is restricted to the appropriate spin direction. $\mu_{xc}[\rho^{\alpha}, \rho^{\beta}]$ is the exchange-correlation energy per electron of a uniform electron gas and the integral (40) sums the contributions from regions of varying density with proper density weighting. The interpretation of Eq. (40) is clear: exchange and correlation at each point are evaluated in terms of what an electron gas with the spin densities found at that point would do. The function μ_{xc} is known in numerical form from highly accurate Monte Carlo calculations and has been fitted to tractable analytical functions by several authors: The fits by Vosko, Wilk, and Nusair [89] and by Perdew and Wang [90] seem to be those most commonly used in the chemical literature. An earlier parameterization by Perdew and Zunger [91] is widely used in the solid state and surface literature.

Although there are small, detailed differences among results from the three parameterizations, their overall behavior is the same, at least for small, light molecules where the great majority of testing has been done. Broadly, the calculated geometries are good, vibrational frequencies and dipole moments also, but atomization energies are too large in magnitude. Molecules are in general predicted to be overbound to the extent that useful thermochemistry does not emerge; errors can be in the range 50 to 200 kJ mol^{-1}. The relatively striking successes have been traced, after much investigation, to the fact that while μ_{xc} provides a poor approximation to the exchange-correlation part of the pair function ρ_2, the spherically averaged exchange-correlation hole it generates satisfies a remarkably large subset of the known constraints. One very physical constraint which it does not satisfy is the asymptotic form of the one-electron potential: far from a molecule, LSDA potentials for both spins fall off exponentially (because the spin density does) and not as $1/r$ as they should.

The failure of the LSDA to get atomization energies right is a clear case of deficient cancellation of spurious self-repulsion; recall the remarks above about the essential role of E_x in this regard. The compact density distributions of atoms (or molecular fragments relative to a large molecule) cause LSDA to have more spurious repulsion for them than for extended systems (e.g. molecules which are more spread out). The result is an artificially large difference between the molecular total energy and the sum of the atomic total energies. There are other

problems with LSDA, for example, the HOMO eigenvalue is a poor approximation to the first ionization potential [Eq. (24) is violated] and van der Waals
binding is not recovered [92,93].

Two earlier versions of LSDA merit brief mention. First is the Slater $X\alpha$ approximation [74]. Instead of μ_{xc} in Eq. (40), it uses a scaled version of μ_x from the
electron gas. That function itself can be shown analytically to go (in the non-
spin-polarized case for simplicity) as $\rho^{1/3}$. The non-spin-polarized $X\alpha$ approximation then is:

$$\mu_{xc}^{X\alpha} = \alpha C \rho^{1/3}(\vec{r}) \tag{41}$$

with C a known constant and α a parameter whose value is 2/3 for the exchange
of the uniform electron gas. Slater and co-workers, however, found that this parameter gave much better results if adjusted (by several different criteria) to values just above 0.7. (For historical reasons the $X\alpha$ approximation sometimes was
called the "Hartree-Fock-Slater" approximation; the name is misleading and
should not be perpetuated.). The $X\alpha$ approximation rarely is used today.

Second is the Hedin-Lundqvist approximation [94] (and the extension to the
spin-polarized case [95]), a simple, clever model of μ_{xc} for the uniform electron
gas. Results from the Hedin-Lundqvist functional are still common in the solid
state and surface literature but the functional seldom has been used for molecules.

Since the functional form of the uniform electron gas μ_{xc} does such a surprisingly good job in real, non-uniform systems, the obvious thought is to pick up
the simplest contributions from non-uniformity by taking the gradient term in
the Taylor series expansion of E_{xc} for the real electron distribution referred to
the uniform one. The resulting gradient expansion approximation fails, basically because it violates many of the constraints on the exchange and correlation
holes. However the idea can be recovered by introducing spin-density gradient
dependencies and forcing as many constraints as possible to be satisfied by in-
genious contrivance of functional forms. It is convenient to introduce the dimensionless, gradient dependent function

$$s(\vec{r}) = \left|\vec{\nabla}\rho\right| / \rho^{4/3} \tag{42}$$

Then a generalized gradient approximation (GGA) looks like Eq. (40) except
for an enhancement factor f_g. The non-spin-polarized versions are equivalent
to:

$$E_{xc}^{GGA} = \int d\vec{r}\rho(\vec{r})\mu_{xc}[\rho(\vec{r})]f_g[\rho(\vec{r}),s(\vec{r})] \tag{43}$$

and the different approximations correspond to differing enhancement factors.

Enhancement factors have been developed via three rather distinct lines of
reasoning. First, consider those generalized gradient approximations that occur
in the context of correcting the low-order gradient expansion. Most of the recent

ones are associated with Perdew and co-workers [90,96–98]. A straightforward expansion yields an exchange-correlation hole $\rho_{xc}(\vec{r}_1, \vec{r}_2)$ [see Eq. (34)] which behaves badly for large values of $t = |\vec{r}_1 - \vec{r}_2|$. The focus of this class of models is to introduce modifications of ρ_{xc} for small and large values of t that preserve the desirable features of the gradient expansion approximation and respect as many of the known constraints on ρ_{xc} as possible. Because these modifications are spatially dependent, geometry-sensitive predictions from the approximations may be affected. Therefore the detailed implementation of the modifications has received much attention leading to several models. All GGA enhancement factors are intricate formulae, so only two will be displayed explicitly. The PW86 exchange functional that multiplies $\mu_x^{LSDA}(\vec{r})$ is

$$f_x^{PW86} = (1 + as^2 + bs^4 + cs^6)^{1/15} \tag{44}$$

where a, b, and c are tabulated constants (note that they are published in different form) [99].

In passing we have introduced the common (but not uniformly followed) nomenclature for these GGA models: the initials of the authors' surnames followed by the year in which the model was published. Thus Eq. (44) is for the Perdew-Wang 1986 exchange functional. There is also a P86 correlation functional [100]. Though these are still in use, from the perspective of satisfying constraints and being accurate, the PW91 exchange and correlation functionals now are accepted as superior [90]. It is important to note that none of the constants in PW86, P86, or PW91 comes from empirical data; all are determined by properties of the exchange-correlation hole (and its system and angular average). This statement also holds for a more recent, formally simpler GGA model that has not as yet been widely used for molecular calculations [98,101,102].

The second class of GGA development is motivated by efforts to reproduce the thermochemistry of small gas-phase molecules and have the correct asymptotic behavior in order to predict the thermochemistry of larger systems. These models do not begin with the uniform electron gas and, in general, do not behave correctly in that case. Perhaps the most active exponent of this route is Becke. His B88 exchange [85] has an enhancement factor of the form:

$$f_x^{B88} = 1 - \frac{\beta s^2}{A(1 + 6\beta s \sinh^{-1}(s))} \tag{45}$$

with A and β constants (note again that the tabulated values have other numerical factors which are omitted here for clarity). The value of β is determined by fitting to exact exchange energies for the rare gas atoms through Rn. Commonly the B88 exchange GGA appears in conjunction with the P86 or PW91 correlation functionals (in which case one has "BP86" or "BP91") or with the LYP correlation functional which we discuss next (whence "BLYP") [103].

To our knowledge, there is only one practical example of constructing a GGA by the third route, which is to take the two-particle reduced density matrix [recall Eq. (29) and discussion leading to it] for a real system and work out the corresponding GGA. The Lee-Yang-Parr (LYP) GGA for correlation originates from such an analysis. These authors started from the approximate two-particle density matrix for the He atom which Colle and Salvetti had suggested [104] and fitted its correlation part to a gradient expansion through second order. Since the system-average hole is all that is needed [recall Eq. (33)], the second-order term can be converted into an additional gradient contribution, so the LYP approximation is a genuine GGA. Though it has been much used lately in the context of hybrid methods (see Sect. 3.3.2), LYP suffers from several deficiencies. At the fundamental level, the LYP approximation is a parameterization of a clever model second-order density matrix which itself is known to be flawed [105]. More practically, the LYP approximation violates a significant number of the known exchange-correlation hole constraints and scaling properties [50].

Salahub and co-workers [106-109] also have constructed and tested functionals based upon the Colle-Salvetti two-particle density matrix. The earlier versions were not gradient-dependent, while the later ones depend upon the Laplacian of the spin densities, hence have the categorical name "LAP". Though they deliver some impressive results on weakly bound systems the LAP models have not become widely used so far.

We conclude this section with some miscellaneous remarks. First, though the GGA models generally improve the calculated molecular energetics, their potentials are quantitatively and qualitatively incorrect locally; see the comparison with exact KS potentials formed to match highly accurate atomic densities from quantum Monte Carlo calculations [110]. Second, the GGA exchange-correlation potentials v_{xc}^{GGA} are functional derivatives with respect to ρ of Eq. (43), so these v_{xc}^{GGA} in fact are strictly local (the potential at a point depends on the density and its gradient at that point alone), hence the terms "semi-local" or "non-local" sometimes used for these approximations are misleading. Third, there are useful tabulations of the various E_{xc} approximations in Refs. [110,111]. The former reference also includes a table of the constraints satisfied by several of the more popular E_{xc}^{GGA} models. Other discussions along the lines of this subsection are also available [112,113].

3.2.2
Hybrid Approximate Functionals

Hybrid methods [22,23] mix the explicitly density-dependent standard KS form of the exchange energy with non-local single-determinant exchange (often somewhat misleadingly called HF exchange). They are motivated by several observations. First, if E_x were to be calculated from the KS orbitals via Eq. (9), the self-repulsion cancellation discussed at various points already would be done correctly. Similarly, this orbital-dependent KS exact exchange term E_x could be subtracted from the adiabatic connection expression Eq. (39) to provide an approach for approximating the correlation term E_c which is a relatively small contribution (in magnitude) to the ground-state energy. Third is the pragmatic observation that errors in calculated binding energies or bond distances for KS methods with approximate functionals, such as those already discussed, and HF methods often are similar in magnitude but have opposite signs. An approach lying between the two methods might therefore benefit from error cancellations. Indeed, in many aspects, hybrid methods are the most accurate DFT procedures available at present.

In hybrid methods the exchange-correlation functional is separated according to

$$E_{xc}^h = \beta E_x\left[\left\{\varphi_i\right\}\right] + (1-\beta)E_x^{appr}[\rho] + E_c^h[\rho]\tag{46}$$

with $0 \leq \beta \leq 1$. The first term is an exchange energy contribution that is calculated exactly via the KS orbitals φ_i according to Eq. (9), while the second term is the remaining part of the exchange energy and is given via an approximate explicit density functional $E_x^{appr}[\rho]$. The third term is a suitable approximate correlation energy functional.

Of course, a question arises: If the exact exchange energy $E_x[\{\varphi_i\}]$ has been calculated, then why use only part of it and leave the rest to an approximation $E_x^{appr}[\rho]$ that introduces errors? Various arguments for this seemingly peculiar procedure appear in the literature. First, Becke [22] argued that a simple trapezoidal approximation to the adiabatic connection integral, Eq. (39), leads to:

$$E_{xc}^{BH} \approx \frac{1}{2}E_x\left[\left\{\varphi_i\right\}\right] + \frac{1}{2}W_{xc}[\rho;\lambda=1] \approx \frac{1}{2}E_x\left[\left\{\varphi_i\right\}\right] + \frac{1}{2}E_x^{appr}[\rho] + \frac{1}{2}E_c^h[\rho]\tag{47}$$

This, the original version of "Becke half and half", is tempting at least insofar as the inadequate cancellation of self-repulsion which has been noted at several points is concerned. Without the factor of one-half the first term in Eq. (47) would cancel that self-repulsion exactly. Somehow a further approximation was introduced in a commercially available code [114]: half and half for only the exchange terms and, after that, in some cases, outright empiricism took over, involving fitting of ansätze, often with rather many parameters. In this mode, one can take advantage of error cancellations between the approximate exchange functional $E_x^{appr}[\rho]$ and the correlation functional $E_c^h[\rho]$.

Indeed, caution is in order if the adiabatic connection formula, Eq. (39), is invoked as motivation for separating exchange into one part that is treated exactly and another that is treated approximately. The exchange contribution to the integrand $W_{xc}[\rho;\lambda]$ [see Eqs. (9) and (39)] is independent of the coupling constant λ. Thus, an integration along the adiabatic connection simply yields E_x and no argument arises for treating E_x in a hybrid fashion. The ultimate argument for reintroducing an approximate exchange energy functional is the desire to take advantage of error cancellation between the approximate E_x^{appr} and $E_c^h = E_c^{appr}$.

So far the standard KS formalism has not been left. If an exchange-correlation potential were obtained via the functional derivative of E_{xc}^h from Eq. (46), then only a special choice of functionals would have been introduced. (As mentioned above, the functional derivative of $E_x[\{\varphi_i\}]$ is not feasible in molecules at present.) It is the hybrid-method equation for the orbitals which leads beyond the realm of the standard KS formalism:

$$\left[(-1/2)\vec{\nabla}^2 + u(\vec{r}) + \beta v_x^{nl} + (1-\beta)v_x(\vec{r}) + v_c^h(\vec{r})\right]\varphi_i = \varepsilon_i\varphi_i. \tag{48}$$

Here, v_x^{nl} is a non-local exchange operator in the form of the HF exchange operator whereas $v_x(\vec{r})$ and $v_c^h(\vec{r})$ are the usual functional derivatives of the energies $E_x^{appr}[\rho]$ and $E_c^h[\rho]$, respectively. The orbital energies emerging from Eq. (48) turn out to fall between those of KS and HF orbitals. (As an aside we note that KS orbitals often have been found to be rather similar to HF orbitals [115] hence hybrid orbitals a fortiori are not expected to be much different.) The corresponding ground state energy

$$\sum_{i=1}^{N}\left\langle\varphi_i\left|(-1/2)\vec{\nabla}^2\right|\varphi_i\right\rangle + U[\rho] + \beta E_x[\{\varphi_i\}] + (1+\beta)E_x^{appr}[\rho]$$
$$+E_c^h[\rho] + \int d\vec{r}\, v_{ext}(\vec{r})\rho(\vec{r}) \tag{49}$$

also resembles the HF and KS expressions, hence is another manifestation of the "hybrid" nature of these procedures.

The success, i.e. the accuracy, of hybrid methods endows them with some justification after the fact. In addition, there is also a sound formal basis for them [116]. The crucial step in a rigorous derivation of hybrid methods is the choice of the model system associated with the real molecular ground state. Recall that the standard KS wave function is the ground state of a non-interacting model system with the same ground-state density as the true molecular ground state. This definition is equivalent to defining the KS wave function as the one which gives the lowest kinetic energy among all those that yield the true molecular ground-state electron density. On the other hand, the HF wave function is defined as the determinant that minimizes the expectation value of $\hat{T} + \hat{V}_{ee} + \hat{V}$. The crucial requirement to justify hybrid schemes is that the model wave function be the determinant that minimizes the expectation value of $\hat{T} + \beta\hat{V}_{ee}$ and yields the true ground-state electron density (Note that the last condition renders this minimization equivalent to that of $\hat{T} + \beta\hat{V}_{ee} + \hat{V}$ since it implies the expectation value of $\hat{V}$ to be constant). The hybrid single-particle equations for the orbitals then can be derived along the same lines as the standard KS equations [116].

The hybrid model wave function differs from both the KS wave function and the HF determinant. As a result the values of the exchange and the correlation energies in a hybrid method differ slightly from those in KS and HF procedures and, furthermore, these energies depend on the value of the parameter β. In practice this β-dependence, as well as the difference from the standard exchange and correlation energies, is more or less neglected. (Sometimes additional semi-empirical parameters in the functionals may account to some extent for these effects; see next paragraph.) For $\beta=0$, the hybrid scheme reduces to the regular KS method, while for $\beta=1$ a modified HF procedure results. It is a HF procedure that, because of the addition of a local correlation potential, yields the correct ground-state electron density and, via Eq. (49), a ground-state energy that contains correlation corrections.

Currently the most popular hybrid method is the semi-empirical B3LYP scheme ("Becke-exchange-3-parameter-Lee-Yang-Parr-correlation"). It owes its origins to a suggestion by Becke [23] for a parameterized hybrid approximation involving the PW correlation functional [90,96,97]. Schematically (brushing over some details) he recommended:

$$E_{xc}^{B3PW} = \beta E_x[\{\varphi_i\}] + (1+\beta)E_x^{LDA}[\rho] + 0.72\Delta E_x^{B88}[\rho] + E_c^{LDA}[\rho] + 0.81\Delta E_c^{PW}[\rho]$$
(50)

with $\beta=0.2$. The third and fifth terms are the shifts of the B88 exchange functional [85] and the PW91 correlation functional, respectively, relative to the corresponding LSDA functionals [90]. The parameters came from fitting to a large set of thermochemical data (on relatively small molecules). With what seems to have been more empiricism, many practitioners adopted Eq. (50) but with the VWN and LYP correlation functionals [89,103] instead of the PW functional, to wit:

$$E_{xc}^{B3LYP} = \beta E_x[\{\varphi_i\}] + (1+\beta)E_x^{LDA}[\rho] + 0.72\Delta E_x^{B88}[\rho] + E_c^{VWN}[\rho] + 0.81\Delta E_c^{LYP}[\rho]$$
(51)

with $\beta=0.2$ as before and $\Delta E_c^{LYP}[\rho]$ the gradient-dependent terms of the LYP correlation functional.

4
Selected Applications

4.1
Calibration and Validation

By now it should be apparent that there is no systematic recipe for building successively more sophisticated DF approximations, in contrast, for example, with perturbation theory. While there is no guarantee that going to the next order in perturbation theory will yield better calculations, there is a certain appeal simply in having the recipe [25]. Whatever the theoretical basis, a prudent approach is to apply a new approximation to a certain class of problems only after explicit testing, i.e. after comparison with experiment or to results of other methods of known accuracy. The unfamiliarity and indirectness of conditions on DF approximations (recall Sect. 3.1 on exchange-correlation hole properties and other constraints) has led to very intense calibration studies of this sort.

Several of the tables that follow include comparisons of DF results with those from various conventional first-principles wave function based quantum chemical methods. In roughly the order of increasing computational cost, MP2 is second-order Møller-Plesset perturbation theory, MCPF is "modified coupled pair functional", QCISD is "quadratic configuration interaction including singles and doubles substitutions", MRCI is "multi-reference configuration interaction", and CCSD(T) is "coupled cluster singles and doubles (perturbative triples)" [25,38,117]. The G2 scheme is a composite one that involves treatment of electron correlation at the MP2 and QCISD(T) levels [35,118] and uses additive empirical corrections to account for remaining deficiencies. The parameterized configuration interaction method (PCI-80) is similar in spirit, but uses one multiplicative parameter based on empirical corrections for the correlation energy [119]; if based on CCSD instead of MP2 results, this method is referred to as PCI-80(CCSD).

Keep in mind the notational scheme by which the approximate exchange and correlation models are designated separately, each by its own letter or acronym. Thus, the de facto LSDA standard will be denoted in the following more explicitly by SVWN [114] (S="Slater" unscaled local density exchange [26] and VWN correlation energy approximation according to Vosko et al. [89]). Furthermore, remember the simple designator B for the B88 gradient corrected exchange approximation [85] (see Sect. 3.2.1). Note that the program Gaussian94/DFT [114] re-interprets the terminology "VWN", which commonly refers to a parameterization based on the electron-gas work of Ceperley and Alder [120] by Vosko et al. [89], to mean an otherwise hardly used LDA variant also described in the latter work. This idiosyncrasy also affects the B3LYP variant of that program [see Eq. (51)].

Early successes of DF methods in optimizing molecular geometries at the LSDA level also included organometallic compounds [11,13,14]. The comparative ease with which transition-metal (TM) carbon bonds were calculated to reasonable accuracy fostered considerable interest in DF methods. Systematic evaluation of small molecules, often diatomics, provided encouraging results in the search for exchange-correlation functionals that yield improved binding energies (e.g. [23,121]). Similarly, systematic testing of GGAs on a wider set of mainly small organic molecules [21] contributed to the popularity of DF methods among computational chemists (see Table 1). These results, obtained with a

Table 1. Comparison of structural and energetic properties (average deviation, AD, and average absolute deviations, AAD, from experiment) for 32 small molecules as calculated with various DF and wave function based methods. Adapted from [21]

Method	Bond distances (pm)		Bond angles (°)		Harmon. vibrational frequencies (cm^{-1})		Atomization energies (kJ mol^{-1})	
	AD	AAD	AD	AAD	AD	AAD	AD	AAD
SVWN	1.4	2.1	−0.74	1.93	−51	75	149	149
BVWN	1.8	1.8	−0.76	1.99	−47	61	0.4	18
BLYP	2.0	2.0	−1.61	2.33	−63	73	4	23
HF	−1.0	2.0	0.11	1.99	165	168	−359	359
MP2	1.0	1.4	−0.87	1.78	69	99	−94	94
QCISD	1.2	1.3	−0.89	1.79	12	42	−120	120

6-31G(d) basis set [122], are of interest from the present point of view as being representative of the accuracy achievable with DF methods for organic ligands. Inspection of Table 1 reveals that, for structural properties of small molecules, DF calculations, either at the LDA or the GGA level, are essentially as accurate as conventional wave function based methods. Calculated vibrational frequencies, even at the LDA level, have errors comparable to those of MP2 and are certainly better than HF results. Note that the properties discussed so far characterize potential energy surfaces in a local fashion around the equilibrium geometry: for those properties, no essential differences were found [21] between LDA and GGA calculations (Table 1). The various methods differed most with regard to atomization energies. Here, gradient corrections, even if applied only for the exchange (BVWN), resulted in significantly improved agreement with experiment (Table 1). Deviations of BLYP results from experiment turned out to be substantially smaller than those obtained with conventional methods at the MP2 or QCISD level.

In view of these structural results it is natural that many efforts to test and validate diverse exchange and correlation models have focused on energetics. A recent example, based on a large thermochemical data set (commonly called the "G2 set" since it was used to parameterize the G2 method [35]), is a study [123] involving 148 molecules: 29 radicals, 35 non-hydrogen systems, 22 hydrocarbons, 47 substituted hydrocarbons, and 15 inorganic hydrides. Even with this breadth, the molecules are rather small and contain only elements from the first and second row of the periodic table, thus only a rather limited number of metals (Li, B, Al). Seven exchange-correlation models, four from first principles plus three hybrid variants that use the B88 exchange GGA, as well as the G2 scheme put forth by those authors were studied. The calculations were carried out with the 6-311+G(3df,2p) basis set [114]. The average absolute deviations (AAD) from experiment for calculated enthalpies (energies of formation plus a pre-

Table 2. Comparison of average absolute deviations (AAD) of enthalpies calculated from four first-principles and three B3 hybrid parameterized DF models relative to experiment, for the five subsets of molecular types which comprise the G2 neutral test set (non-hydrogen molecules, hydrocarbons, substituted hydrocarbons, radicals, and inorganic hydrides). Adapted from [123]. All values in kJ mol^{-1}

	35 Non-H	22 H-C	47 Subst. H-C	29 Radicals	15 Inorg. H
SVWN	308	559	520	228	141
BP86	69	108	112	66	34
BPW91	51	20	33	27	18
BLYP	43	34	26	25	13
B3P86	33	129	107	57	33
B3PW91	22	17	12	13	8
B3LYP	22	12	9	12	8
G2	11	5	8	6	4

Table 3. Comparison of average absolute (AAD) and maximum deviations (MD) of barrier heights (in kJ mol^{-1}) for 12 small-molecule organic reactions for two conventional methods, HF and MP2, and the BLYP and B3PW91 density functionals. After [124,125]

	AAD	MD
HF	57	128
MP2	41	120
BLYP	25	92
B3PW91	15	54

scribed thermal correction; for details see Ref. [123]) are given for the G2 subsets in Table 2. Not unsurprisingly, the two newer three-parameter functionals reflect the fact that the parameters are fitted to this kind of molecule. If the 15 kJ mol^{-1} variation is taken as a reasonably stringent measure of "chemical accuracy", a notable feature of Table 2 is that even first-principles GGA models are close. However, the enthalpies from the G2 scheme exhibit deviations from experiment which are essentially only half as large as those of the best DF variant B3LYP.

Table 3 gives a glimpse of the quality of calculated activation barrier heights that currently is attainable on typical small-molecule organic reactions (six each closed shell and radical) [124,125]. (Note that what is called the adiabatic connection method, "ACM", in Ref. [124] is B3PW91.) As Table 3 suggests, barrier heights are a serious problem for all relatively affordable methods. Among DF models, the LSDA functionals are quite unreliable, as an early study of various organic and main-group organometallic reactions showed [126]. However, with the GGA(BP86) model, results of a quality comparable to MP2 were obtained. For a recent review of the performance of DF methods (including hybrid variants) for the calculation of barrier heights see Ref. [127]. These studies [125,126] support the general experience that barrier heights calculated by DF approximations tend to be lower than those from conventional methods; B3LYP shows generally better agreement with experiment than BLYP. A particularly spectacular failure in this context is that LSDA (SVWN) predicts H$_3$ to be a stable species instead of a saddle point of the exchange reaction H$_2$+H→H+H$_2$. This failure has been traced to insufficient self-interaction correction [128].

Before turning to DF results for TM organometallic compounds in the following section, we discuss the prediction of relationships among low-lying excited states of isolated TM atoms. As mentioned in the Introduction, these present particular challenges to all quantum chemical methods. An accurate description of the energetics of low-lying configurations is crucial for describing the ability of a TM atom to undergo configuration change and re-hybridization under the influence of approaching ligands. There are several studies which bear on the problem [13,14,129,130]. Recently Siegbahn and coworkers have studied this topic and its consequences in detail with regard to the performance of various methods when describing reactions of TM compounds [41,131]. For conven-

Table 4. Energy difference between the ground and first excited states of first-row TM cations. After [130,131]. All values in kJ mol^{-1}

Cation	SVWN	BP86	B3LYP	MCPF	PCI-80	Exp.
Sc	47	45	35	92	91	58
Ti	−38	−32	−5	43	37	10
V	103	94	31	3	15	32
Cr	170	174	137	125	139	147
Mn	124	121	142	237	207	175
Fe	20	5	11	80	50	24
Co	70	82	37	−16	18	41
Ni	179	169	81	46	85	104

ience Table 4 displays a few illustrative results from a recent investigation on the monocations of the first-row TM atoms Sc to Ni [131]. From a computational point of view these elements are very challenging because of the compact form of the 3d orbitals. An accurate description of the relative energies of low lying configurations is particularly crucial therefore. In passing we note that, unlike anions, cations do not suffer from a well-documented and well-understood problem with present-day pure DF approximations, to wit, destabilization by improper self-repulsion [132]. Even so, the inadequacy of cancellation in such approximations yields poor calculated excitation intervals. Hence the situation communicated by Table 4 represents about the best that can be achieved at present. The details of the transitions are not important here. What is significant is that MCPF does extremely poorly, while even the early first-principles GGA functional BP86 does quite well, except for the late TMs. The B3LYP semiempirical DF model is as good if not better than the semi-empirical PCI-80 scheme [131] in representing the general trends across the whole series. Unfortunately for the chemistry of certain elements, each method exhibits noticeable failures: every model except PCI-80 actually gets one excitation interval upside down, i.e. the wrong ground state, and errors range from zero to a factor of 2 (B3LYP) to 3 (PCI-80).

The dissociation TMH^{+}→TM^{+}+H provides the simplest test case for calculating a TM-ligand bond strength [130,131]. Similar to the results of splitting between the two lowest lying configurations of the cations TM^{+}, the accuracy of the GGA functional BP86 results is also quite satisfactory for the early TMs (about 20 kJ mol^{-1}), but less so for the late TMs (up to about 55 kJ mol^{-1}) [130]. For the B3LYP semiempirical DF and conventional MCPF and PCI-80 methods [131], the absolute deviations of the calculated bond dissociation energy (including vibrational corrections) from experiment, averaged over the first-row TM series, are 20, 25, and 8 kJ mol^{-1}, respectively. Note that the average uncertainty of the experimental results is almost 9 kJ mol^{-1} [131]. The B3LYP bond dissociation energies are consistently too large, but of acceptable accuracy, although in this case not as reliable as the results of the PCI-80 method.

4.2
Organometallics – Broad View

We focus now on organometallic compounds of TMs and begin by displaying a sample of the available comparisons with experiment for such basic quantities as binding energies, bond lengths, and vibrational frequencies.

Table 5 gives the LSDA (SVWN) and GGA (BP86) comparison for metal-carbonyl bond dissociation energies (BDE) reported by Li et al. [138]. The major improvement over LSDA which is afforded by even a relatively simple GGA is evident, as is the quality of the results relative to experiment and to the more demanding conventional methods. (Note that earlier results by Ziegler and coworkers [11] were for incompletely optimized species, but even so were attention-getting; they have been superceded by the results summarized here, which are fully geometry optimized.)

The overbinding characteristic of LSDA is quite evident, as is the major improvement afforded by even a simple, relatively early GGA. Clearly the Cr species is a tough case for all currently available methods, but neither cost nor formal sophistication are a portent of success. Similar GGA results have been reported by Jonas and Thiel [139]. Siegbahn and coworkers have calculated the successive BDEs of all ligands of $Ni(CO)_4$ at the B3LYP level [131].

Fournier and Pápai have carried out an extensive study of mono-ligand TM complexes [140] from which we present data for Cu and Ni systems (see Table 6). BDEs of monoligand TM systems afford a particularly stringent test of a computational procedure since, in general, near-degeneracy effects are very important. As an aside as well as a note to Table 6, their work is an example of a fairly stand-

Table 5. First metal–carbonyl bond dissociation energies for various DF and conventional methods (see text for notation) as well as from experiment. Adapted from [140]. All values in kJ mol^{-1}

	$Ni(CO)_4$	$Fe(CO)_5$	$Cr(CO)_6$	$Mo(CO)_6$	$W(CO)_6$
SVWN	188	276	260	220	202
BP86	120	187	192	160	162
BP86+Rel.[a]	125	191	193	166	183
MP2[b]	NA[c]	NA[c]	243	193	230
MCPF	100[d]	163±21[e]	NA[c]	NA[c]	NA[c]
CCSD(T)	125[f]	NA[c]	192[b]	169[b]	201[b]
Exp.[g]	105±8	176	154±8	169±8	192±8

[a] Scalar-relativistic corrections included.
[b] Ref. [133].
[c] Not available.
[d] Ref. [134].
[e] Ref. [135].
[f] Ref. [136].
[g] Ref. [137].

Table 6. Comparison of GGA (BP86) and experimental dissociation energies for Cu and Ni monoligand systems. Adapted from [140]. All values in kJ mol^{-1}

	Calc.	Exp.
CuCO	71	25
$Cu_2(CO)$	109[a]	>105
$Cu(C_2H_4)$	29[a]	25
$Cu_2(C_2H_4)$	79[a]	>84
$Ni(C_2H_2)$	192	192±25
$Ni(C_2H_4)$	163	146
$Ni(H_2CO)$	134	NA[b]
$Ni[(CH_3)_2CO]$	130[a]	134

[a] Geometry optimized with SVWN.
[b] Not available.

ard procedure for reducing computational effort, namely, determination of the molecular geometry via LSDA (SVWN) and subsequent binding energy calculation at that geometry via GGA (either BP86 or PW86 in their case). Different GGA variants often have been found to exhibit similar binding energies of ligands to TM atoms, with BP86 or BP91 values slightly larger than BLYP (e.g. Ref. [141]).

So far, we have mainly discussed DF test calculations for carbonyl and π-bonding ligands of transition metal centers. Methyl and methylene groups afford different types of metal-ligand bonding which are crucial for much of TM organometallics. Siegbahn and coworkers calculated the BDEs for first-row transition-metal species TM– CH_3^+ and TM = CH_2^+ with B3LYP and various conventional methods [131]; the results are compiled in Tables 7 and 8. These systems provide rather stringent tests for any quantum chemical method since strong near-degeneracy effects may occur due to the fact that the TM centers are coordinatively unsaturated. Binding energies were determined with large all-electron basis sets in single-point fashion at geometries determined either by B3LYP in conjunction with effective core potentials (ECP) [4] (TM–CH_3^+) or taken from previous studies (TM=CH_2^+) [142]. Investigations of similar thrust can be found in Refs. [143–145]. The deviations in BDE of TM-CH_3^+ are the same order of magnitude as for TM–H$^+$. The similarity is not too surprising since in both cases a single bond between the metal center and the ligand is broken. Nevertheless the deviations from experiment are slightly larger for the methyl derivatives than for the hydrides with the exception of PCI-80 (see Sect. 4.2). This outcome may be rationalized by the fact that the methyl group also has electrons in closed shells that can interact in a repulsive manner with the electrons of the metal 3d shell [131]. The electron correlation involved is described differently by the several methods and B3LYP turns out to be favored over some (very time-consuming) conventional methods like CCSD. In Fig. 1 the deviations of calculated re-

Table 7. Metal–ligand bond dissociation energies of first-row TM–CH$_3^+$ systems. Also shown is the average absolute deviation from experiment, AAD. Adapted from [131]. All values in kJ mol^{-1}

	B3LYP	MCPF	PCI-80	CCSD	CCSD(T)	PCI-80 (CCSD)	Exp.
Sc	254	201	232	199	209	230	233
Ti	246	181	224	NA[a]	NA[a]	NA[a]	214
V	209	154	188	154	167	188	192
Cr	149	83	122	82	99	121	110
Mn	210	157	193	152	164	187	205
Fe	261	198	236	192	206	230	228
Co	223	166	204	163	178	201	203
Ni	206	134	175	141	157	184	187
AAD	23	37	8	39	26	6	

[a] Not available.

Table 8. Metal–ligand bond dissociation energies of first-row TM=CH$_2^+$ systems. Also shown is the average absolute deviation from experiment, AAD. Adapted from [131]. All values in kJ mol^{-1}

	B3LYP	MCPF	PCI-80	CCSD	CCSD(T)	PCI-80 (CCSD)	Exp.
Sc	351	296	367	285	316	353	372
Ti	355	298	385	274	310	355	380
V	318	255	354	223	262	314	326
Cr	239	159	277	117	165	223	216
Mn	287	182	295	126	177	226	286
Fe	315	231	331	191	239	276	341
Co	325	226	318	199	244	284	318
Ni	297	218	308	194	235	278	306
AAD	15	85	15	117	74	31	

sults from experiment are compared for several methods; all B3LYP BDEs are too large compared to experiment. Again, the Cr complex is noticeably difficult to calculate with all methods except PCI-80 (cf. Table 5).

The bonding between a carbene ligand and a TM center exhibits both σ and π components. Thus, such bonding turns out to be significantly more difficult to describe than the single bond of a methyl ligand to a TM, as the results of all conventional quantum chemical methods for the BDE of TM$=$CH$_2^+$ show (Table 8). In fact, these AAD values are at least twice those that were obtained for TM–CH$_3^+$ (cf. Table 7 and Fig. 1). It is interesting that B3LYP performs in excellent fashion, on par with PCI-80. In view of the computational costs involved, it is so-bering to note that the AAD of B3LYP is about a factor of eight smaller than the corresponding CCSD value and even about a factor of five smaller than that of CCSD(T).

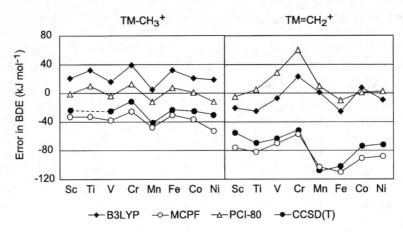

Fig. 1. Errors of calculated dissociation energies of TM–CH$_3^+$ and TM=CH$_2^+$ bonds for various methods. Adapted from [131]

Turning to bond lengths, Table 9 gives comparisons for five carbonyls, again from Ziegler and co-workers [138]. (Closely related results are in Refs. [139,149,150].) Among the striking features exhibited in Table 9 is the dramatic improvement in bond lengths from LSDA to GGA. Note that the lack of success with bond energy for the Cr species (Table 5) does not carry over to bond length. The data in Table 9 show remarkable failure of conventional approaches to do better than DF methods; as often as not the results are worse. Not unexpectedly, scalar-relativistic corrections turn out to be very small for all compounds of first- and second-row TMs, but a significant relativistic shortening of the TM–carbon bond is calculated for W(CO)$_6$.

Finally, we discuss vibrational frequencies of TM organometallic compounds. Infra-red (IR) and Raman spectra of TM carbonyl compounds are experimentally well characterized, thus many comparative computational studies are available (e.g. Refs. [133,138,139,149,151,152]).

We start with a detailed comparison of all the harmonic vibrational frequencies of Cr(CO)$_6$ as calculated by various GGA methods (see Table 10). The computed values are displayed as deviations from experimental frequencies. For the C–O stretching modes (v_1, v_3, and v_6) the DF results are lower than the experimental values; indeed, deviations are up to 50 cm^{-1} relative to harmonic estimates [157]. On the other hand, calculated M–C stretching frequencies are higher than experiment. With the exception of v_7 the deviations for all valence angle deformation modes are below 10 cm^{-1}. All computational strategies compared in Table 10 perform equally well as manifest from the AAD of all calculated frequencies. The present molecule is not an exceptional case: GGA methods in general have been found to perform very well for the calculation of vibrational frequencies [133,138,139,149]. The results of Table 10 demonstrate that the accuracy does not vary much with the quality of the basis sets and or the GGA variant

Table 9. TM–carbon bond lengths in five carbonyls[a] from LSDA and GGA(BP86) calculations compared to results of conventional methods and experiment. Adapted from [138]. All values in pm

	$Ni(CO)_4$	$Fe(CO)_5$	$Cr(CO)_6$	$MO(CO)_6$	$W(CO)_6$
SVWN[b]	177.9	176.9; 178.9	186.6	203.5	206.0
BP86[b]	183.0	181.9; 181.6	191.0	207.7	211.6
BP86+Rel.[c]	182.9	181.7; 181.3	191.0	207.6	204.9
MP2	187.3[d]	NA[e]	188.3[f]	206.6[f]	205.4[f]
MRCI[g]	NA[e]	179.8; 183.5	NA[e]	NA[e]	NA[e]
MCPF	188.3[h]	187.8; 184.7[i]	NA[e]	NA[e]	NA[e]
CCSD(T)	183.1[j]	NA[e]	193.9[k]	NA[e]	NA[e]
Exp.[l]	183.8	180.7; 182.7	191.8	206.3	205.8

[a] Symmetries: C_{2v} for $Ni(CO)_4$, D_{3h} for $Fe(CO)_5$ (equatorial bonds listed first, axial second), O_h for $M(CO)_6$.
[b] Ref. [138].
[c] Scalar-relativistic corrections included.
[d] Ref. [146].
[e] Not available.
[f] Ref. [133].
[g] Ref. [147].
[h] Ref. [134].
[i] Ref. [135].
[j] Ref. [136].
[k] Ref. [148].
[l] See Ref. [138].

employed. This opens the possibility of saving computational effort if highly accurate frequencies are not so important, e.g. when checking the eigenfrequencies of a transition state and for zero point corrections to the reaction energies.

Fournier and Pápai [140] also have reported extensive calculations focused on IR spectra and intensities of TM complexes with single ligands in which they compare LSDA and GGA results to experiment. They studied all first-row TM and some second-row mono-carbonyls, as well as complexes with water, ethylene, methanol and ammonia and were able to conclude that the triplet ground state of these complexes is the likely assignment. Table 11 shows their results for harmonic frequencies from GGA (BP86) for a small ethylene complex (both normal and deuterated); we exhibit only those of their entries for which comparison with experiment was made. With this overview for perspective, we next examine a specific case in detail.

Table 10. Vibrational frequencies ν and deviations $\Delta\nu$ of GGA results for $Cr(CO)_6$. Adapted from [139]. Also shown is the average absolute deviation, AAD, from experiment. All values in cm^{-1}

| | | | ν, Exp.[a] | $\Delta\nu$, BP86 | | | $\Delta\nu$, PW91 |
				AE[b]	ECP[c]	TZ/DZP[d]	DNB[e]
A_{1g}	ν_1	[CO]	2119	−26	−19	−31	−9
A_{1g}	ν_2	[MC]	379	21	22	6	30
E_g	ν_3	[CO]	2027	−23	−10	−29	−9
E_g	ν_4	[MC]	391	16	16	3	25
T_{1g}	ν_5	[δMCO]	364	1	−6	−4	1
T_{1u}	ν_6	[CO]	2000	−16	−1	−22	−1
T_{1u}	ν_7	[δMCO]	668	22	23	19	26
T_{1u}	ν_8	[MC]	441	20	17	6	31
T_{1u}	ν_9	[δCMC]	97	1	6	−3	5
T_{2g}	ν_{10}	[δMCO]	532	2	1	2	−2
T_{2g}	ν_{11}	[δCMC]	90	−1	1	−1	5
T_{2u}	ν_{12}	[δMCO]	511	8	10	1	1
T_{2u}	ν_{13}	[δCMC]	68	−7	−6	−2	−3
AAD				13	11	10	11

[a] Ref. [157]. Harmonic estimates are ν_1: 2139, ν_3: 2045 and ν_6: 2044 cm^{-1}.
[b] Augmented (p- and d-exponents) all-electron basis set [153] for Cr, together with a triple-ζ basis set [154] for C and O.
[c] Quasi-relativistic ECPs [155] for Cr with 6-31G(d) basis set for C and O.
[d] Triple-ζ basis set for Cr and polarized double-ζ basis set for C and O. Values taken from [151].
[e] Double numerical basis. Values taken from [156].

Table 11. Calculated harmonic frequencies ν of $Ni(C_2H_4)$ and shifts $\Delta\nu$ upon C_2H_4/C_2D_4 isotopic substitution in comparison with experiment[a]. Adapted from [140]. All values in cm^{-1}

| | ν, $Ni(C_2H_4)$ | | $\Delta\nu$, $Ni(C_2D_4)$ | |
	BP86	Exp.[b]	BP86	Exp[b]
$2a_1$ CC stretch	1461	1465	190	196
$3a_1$ CH$_2$ scissors	1158	1156	246	230
$5a_1$ NiC stretch	539	498	37	12
$3b_2$ CH$_2$ wag	923	901	162	286

[a] Only those frequencies are displayed for which comparison with experiment was made.
[b] Ref. [158].

4.3
Organometallics – Case Study of Transition-Metal-Catalyzed Oxygen Transfer Reactions

This section gives a detailed survey of the use of the theoretical and computational methodology for resolution of a difficult chemical problem. The topic is reaction mechanisms which are important in catalysis. Kinetic properties such as transition states are the primary determinative factors for chemical reactivity. Though difficult to explore experimentally, in principle they are relatively easy to calculate. Since we will deal mainly with TM compounds, two aspects of typical computational treatments of such systems must be mentioned.

First, it is common to use effective core potentials (ECP). These are local potentials which substitute, as far as the valence electrons are concerned, for the core electrons. Their advantage is that the tightly bound core electrons no longer have to be described by the basis set used to expand the KS orbitals, hence the basis can be both smaller and better adapted to the representation of the valence orbitals. As an aside we note that for heavy elements ECPs also include relativistic effects in a convenient fashion. Alternatively, one has to resort to a relativistic DF method which implies the solution of a four-component Dirac-KS equation (for an overview of relativistic DFT and an efficient formalism for solving the Dirac-KS problem in molecules, see Ref. [159]). Basis set choice is the second technical issue. A useful starting point is the Los Alamos small core ECP set as implemented in the standard compilation LanL2DZ [160] in Gaussian94 [114], but for quantitative results one has to modify the basis sets. A detailed discussion can be found in Ref. [5].

We will discuss different computational strategies as applied to two important oxygen transfer reactions to olefin substrates: dihydroxylation [161] catalyzed by osmiumtetroxide OsO_4 and epoxidation [162,163] mediated by methyltrioxorhenium CH_3ReO_3 (MTO) in the presence of hydrogen peroxide [164]. Pidun et al. [165], Dapprich et al. [166], Torrent et al. [167] and DelMonte et al. [168] investigated mechanistic details of the former reaction while the latter was studied by Gisdakis et al. [169] and Wu and Sun [170].

4.3.1
Olefin Dihydroxylation with OsO_4/H_2O_2

Addition of an olefin to an OsO_4 moiety (**1A**, see Scheme 1) has attracted much study in the last few years. Sharpless and coworkers [161] were able to develop a stereoselective variant of this reaction by adding a chiral base to the catalyst. In that context a new mechanism was suggested, in which the olefin first attacks an Os=O bond in a [2+2] fashion to give an intermediate osmaoxetane, which should then rearrange to an osmate ester (**2A**, see Scheme 1). This latter product could also be formed by a concerted [2+3] cycloaddition of the olefin to the O= Os=O unit, as suggested by Criegee et al. [171].

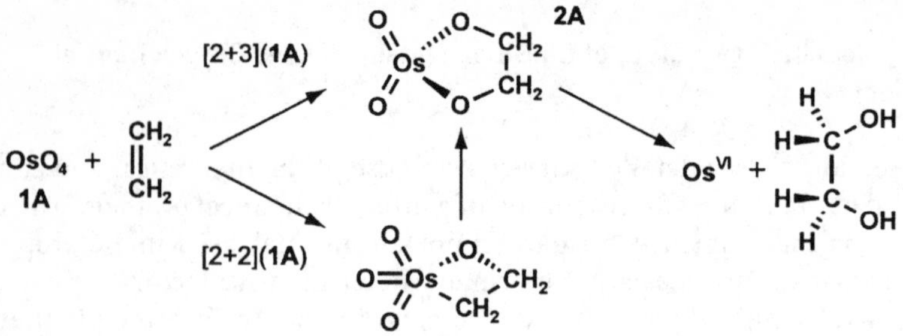

Scheme 1

Sharpless' suggestions were based on experimental findings which indicate a two-step mechanism. Intermediate formation of an osmaoxetane can also explain the stereochemical induction of the chiral ligand found experimentally [161]. Although these suggestions are reasonable, there are serious theoretical objections against a [2+2] mechanism. DF calculations for the bare complex **1A** (or the corresponding complex **1B** with an additional NH_3 ligand, not shown in Scheme 1) showed that the formation of the intermediate osmaoxetane is kinetically blocked by a very high barrier for the [2+2] cycloaddition [165-168]. Despite the differences in models and methods, calculations from four groups agree that the [2+2] pathway barrier is in the range of 165 to 185 kJ mol^{-1}, while the [2+3] pathway barrier lies in the range of 8 to 40 kJ mol^{-1}. The calculated value of the barrier height depends on various factors (see Table 12): (1) whether a base is coordinated to the metal center, (2) the method used (B3LYP, BP or CCSD(T); recall acronyms from preceding Section), and (3) the quality of the basis set. Before analyzing the results, a summary of important computational details is appropriate.

– Study I: Pidun et al. [165] optimized all structures with B3LYP, then calculated the energy at the optimum structures (a "single point" calculation) with the CCSD(T) method. For Os they used a small core ECP with a (441/2111/21) basis set. All other atoms were treated with a 6-31G(d) basis set. The same basis sets were used for the CCSD(T) step, except for the H atoms, which were treated with an STO-3G basis set.

– Study II: Dapprich et al. [166] also optimized all structures with B3LYP, followed by a single-point CCSD(T) energy calculation but used the (smaller) LanL2DZ basis set for all atoms and all calculations.

– Study III: Torrent et al. [167] performed all geometry optimizations within the local density approximation LDA (SVWN) followed by a single point energy calculation with the gradient corrected BP86 exchange-correlation functional. Instead of an ECP, a frozen core approximation was used. The outermost atomic valence orbitals (for Os, outermost plus one) were treated with a triple-ζ STO basis set. A quasi-relativistic correction was included for the total energies.

– Study IV: DelMonte et al. [168] used the same optimization strategy as I, but employed a (341/321/21) basis set for Os.

Table 12. Comparison of activation barriers (in kJ mol^{-1}) as obtained in various studies (see text for details). Designation of the various transition states TS as in Scheme 1

TS	I[a]		II[b]	III[c]	IV[d]
	B3LYP	CCSD(T)	CCSD(T)	BP86	B3LYP
[2+2](1A)	184	187	181	166	–
[2+3](1A)	21	40	8	6	–
[2+2](1B)	185	–	211	164	185
[2+3](1B)	18	–	6	3	13

[a]Ref. [165].
[b]Ref. [166].
[c]Ref. [167].
[d]Ref. [168].

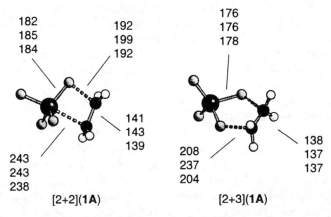

Fig. 2. Structures and bond lengths (in pm) of the dihydroxylation transition states [2+2](1A) and [2+3](1A) as obtained in computational studies I to III (top to bottom; see text)

The data from study I for B3LYP and CCSD(T) show that energetically these two methods compare very well for the calculated barriers. The difference for the transition state [2+2](1A) is very small (3 kJ mol^{-1}) while it is larger for [2+3](1A), 19 kJ mol^{-1}. Another discrepancy attracts attention: the CCSD(T) energies from studies I and II for the transition state [2+2](1A) differ by about 6 kJ mol^{-1} while for [2+3](1A) the deviation is 32 kJ mol^{-1}. Results from Table 12 are in line with the general experience that reaction barriers calculated with DF methods tend to be too low (recall Table 3). With the exception of study II (see below), values from GGA are consistently lower than those of hybrid methods which in turn lie below the CCSD(T) values.

A review of the optimized transition state structures (see Fig. 2) makes it apparent that the energetic differences arise from different optimized geometries.

Those differences, in turn, are caused by the lack of polarization functions in study II. It seems that the reaction coordinate with the larger barrier, the [2+2] cycloaddition, is not as sensitive to the inclusion of polarization functions in the basis set as the "softer" reaction path, the [2+3] cycloaddition. Inspection of Fig. 2 shows that for the transition state [2+2](**1A**) the agreement of geometries calculated in Studies I and II is good while for the latter calculation the C–O bond length is about 30 pm longer for transition state [2+3](**1A**). These facts may be rationalized by the differences in curvature of the potential energy surfaces at the transition state locations. The potential energy surface at the low-energy transition state is much flatter than at the higher lying one, thus a small change in the parameters of the calculation leads to noticeable effects in the optimized structure. A second result evident from Fig. 2 is the well-known fact that LSDA tends to overbinding: the structures of study III optimized with SVWN show rather shorter bond lengths than those from B3LYP (Study I).

Kinetic isotope effects allow a check of calculated and experimental properties of a transition state. For the asymmetric dihydroxylation with OsO_4, such a comparison was carried out in study IV [168]. The kinetic isotope effect for the [2+3] cycloaddition was found to agree very well with experiment while for the [2+2] cycloaddition it did not, thus providing further evidence for the conclusion on the mechanism that is based on the calculated activation barriers.

4.3.2
Olefin Epoxidation with CH_3ReO_3/H_2O_2

Extensive experimental work has been devoted to methyl trioxorhenium (MTO) which has proven to be a highly efficient olefin epoxidation catalyst in the presence of hydrogen peroxide [164]. MTO reacts with H_2O_2 resulting in mono- and bisperoxo compounds. An additional aquo ligand has been found to stabilize the latter complex [172]. Herrmann et al. [172] and Espenson et al. [173] have proposed reaction mechanisms for epoxidation by MTO-related complexes involving differently oxygenated and hydrated forms of these catalysts.

The techniques used in a recent DF study [169] of this reaction represent quite a high level of calculation. Geometries were optimized using the B3LYP exchange-correlation functional with effective core potentials and double-ζ basis sets, LanL2DZ [160] on Re and 6-311G(d,p) on H, C, and O, followed by single-point energy calculations, with a (441/2111/21/11) decontracted LanL2DZ basis set on Re [5] with two f-type exponents added on Re. The enthalpy values ($T= 298.15$ K) were calculated using vibrational frequencies obtained with LanL2DZ basis sets.

We start by considering various rhenium oxo and peroxo complexes (Fig. 3): MTO (**3**) as well as the corresponding mono- (**4**) and bisperoxo (**5**) complexes, each of them in free (**A**) and monohydrated (**B**) form. The water ligand of complexes **B** was added in the *cis* position to the methyl group, as suggested by the X-ray structure for the bis-peroxo complex [172]. As expected for a Re center with strong Lewis acidity [174], hydration leads to significant stabilization,

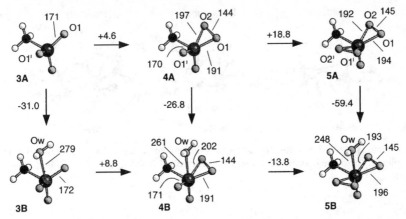

Fig. 3. Methyl trioxorhenium (MTO, **3A**) and the corresponding mono- (**4A**) and bisperoxo complexes (**5A**) as well as the monohydrated complexes **3B**, **4B**, and **5B**. Bond lengths in pm, enthalpy changes of peroxidation (*rows*) and hydration (*columns*) in kJ mol^{-1}

yielding an important contribution to the energy of MTO peroxidation (Fig. 3). Agreement between calculated and experimental [172] geometries of **3B** is very satisfactory, with largest bond length deviations in general about 4 pm. The metal-water bond length is an exception (Re–O$_w$: exp. 225 pm [172], calculated 248 pm; Fig. 3). This deviation can be rationalized by a co-crystallized ether molecule [172]. Modeling of that crystal environmental effect by two dimethyl ether molecules (which form hydrogen bonds with the coordinated water) reduces the calculated value of Re–O$_w$ to 234 pm. The peroxo groups are asymmetrically bound to the rhenium center: the tilting of the peroxo ligands relative to the "back" and "front" side (close to and far from the methyl ligand, respectively) is opposite in complexes **4** and **5** (cf. bond lengths Re–O1 and Re–O2, Fig. 3).

Since MTO is catalytically active in the presence of H$_2$O$_2$ [172,173], possibly active species in the olefin epoxidation reaction are the unsolvated (**4A**, **5A**) and mono-solvated peroxo complexes (**4B**, **5B**). Three types of reaction pathways were studied computationally [169]: (1) Direct olefin transfer [175] occurs when the olefin double bond attacks a peroxo group, either from the front (O1, "spiro" **S**) or back (O2, "back-spiro" **BS**); the calculations show that the olefin approaches the peroxo group in a spiro, rather than a planar fashion (see Fig. 4). (2) Insertion (**I**) via a [2+2]-like arrangement [176] leads to a five-membered ring intermediate involving the metal center and a peroxo ligand (Fig. 4). The epoxide is extruded via a second transition state which was not studied further because the corresponding first step is energetically unfavorable compared to the spiro processes (see below). (3) Oxygen transfer also may occur from a hydroperoxo species which appears as an intermediate during the peroxidation of MTO (or its mono-peroxo derivative). A hydroperoxo complex may be derived formally by proton transfer from the water ligand (in **4B** or **5B**) to a peroxo group. There is

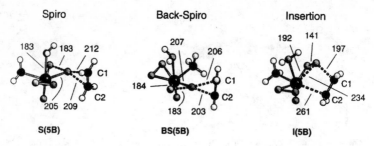

Fig. 4. Various transition state structures for ethylene epoxidation by MTO-derived peroxo complexes: spiro, back-spiro, and insertion processes involving complex **5B**. Bond lengths in pm

Table 13. Calculated activation enthalpies $\Delta H^{\#}_{298}$ (in kJ mol^{-1}) for olefin epoxidation via spiro **S**, back-spiro **BS** and insertion **I** processes, starting from various MTO-related complexes (see Fig. 5)

	B3LYP			BLYP[a]	
	S	BS	I	S	BS
4A	77	79	158	51	55
4B	–[b]	69	–[b]	56	45
5A	54	82	111	28	58
5B	70	100	165	44	69

[a] Adapted from [170].
[b] No transition state localized since the water ligand is expelled during the optimization.

experimental [177] as well as theoretical [178] evidence that such hydroperoxo species are active in olefin epoxidation by molybdenum complexes. However, in the system MTO/H_2O_2, calculations have shown that hydroperoxo species are not competitive with MTO-related peroxo complexes [169]; thus we refrain from further discussion of them.

Compared to calculated transition states of organic epoxidation reactions [179], **S** and **BS** transition states of MTO-derived peroxo complexes feature an almost concerted approach of the olefin double bond to the active oxygen center where the attacked peroxo bond O1-O2 is significantly lengthened, by about 40 pm, while the Re-O bond of the oxygen center to be transferred (**S**: O1, **BS**: O2) is elongated by about 10 pm only (see Fig. 4). During insertion the attacked Re-O bond is more elongated than in the spiro processes, while the corresponding O-O bond is somewhat contracted.

The calculated activation enthalpies $\Delta H^{\#}_{298}$ for ethylene epoxidation are compared in Table 13. We first discuss the results obtained with the B3LYP hybrid method [169]. Obviously, for each complex insertion, **I** exhibits a noticeably higher barrier than both types of spiro approaches, S and BS. Comparing the two variants of the direct transfer for each starting complex, we note that for 5 a spiro attack from the "front" is clearly favored over a back-spiro process, while the latter mechanism is competitive for 4A and is the only one found for 4B. In general the water ligand enhances the barriers for all three types of mechanisms, with the exception of transition state BS(4B) [cf. BS(4A/B); we note the starting complex of a transition state in parentheses]. Hydration leads to a decreased charge on the metal center that in turn causes the oxygen centers to become more negative. The increase of the barriers may be rationalized if one recalls that epoxidation is a nucleophilic attack of the olefin on an oxygen atom [163].

The low barrier of the transition state BS(4B) correlates with a significant elongation of the attacked bond Re–O2 (by 5 pm). Transition state S(5A) has the lowest calculated barrier, but complex 5A is significantly less stable than 5B (by 59 kJ mol^{-1}, Fig. 3). Thus, we conclude that epoxidation starting from 5 is likely to proceed via transition state S(5B). The "back-spiro" transition state BS(4B) has essentially the same activation energy as S(5B). Since the mono-peroxo complex 4B is only slightly less stable (by 14 kJ mol^{-1}) than the bisperoxo complex 5B, processes S(5B) and BS(4B) are quite competitive with preferences possibly determined by entropy effects or experimental conditions. The calculated B3LYP activation enthalpies are comparable with the measured value [173], 42.7±1.7 kJ mol^{-1}, for epoxidation of 4-methoxystyrene with a bisperoxo complex.

Comparable work was published very recently by Wu and Sun [170]. In it, smaller basis sets, equivalent to 6-31G(d) (see Ref. [170] for details) and the BLYP GGA exchange-correlation functional were used. As a result, the absolute height of the barriers changes (they are consistently lower than with B3LYP), while the relative ordering of the various processes remains the same. An exception is mechanism S(4B), for which the higher level of calculation [169] finds the water ligand to be expelled during the reaction while it sticks to the Re center in the calculation [170] which used a smaller basis set. Thus, this sticking can be viewed as a basis set error as confirmed by a comparison of results obtained for different basis set sizes [178]: the metal center uses diffuse basis functions from the ligand elements to saturate its own, insufficient basis set (and vice versa) and therefore a kind of virtual attractive interaction results. A second and far more dramatic difference is in the structures of the optimized transition states. The BLYP model [170] predicts a more ionic reaction mechanism than the B3LYP model [169]. This difference leads to a non-symmetrical approach of the olefin to the oxygen center. Singleton et al. showed in a comparative study that experimental and calculated kinetic isotope effects fit very well for B3LYP, while they show bigger deviations for other methods such as MP2 [179]. Their conclusion is that the more concerted mechanism, which is favored by B3LYP, fits better with the experiment.

In conclusion, the calculated results for olefin dihydroxylation and epoxidation by oxo and peroxo complexes discussed in this section confirm that today hybrid DF methods provide an efficient tool for studying TM organometallics and their reactivity. Hybrid DF methods rival the best conventional wave function based methods in accuracy, yet require only a fraction of their computational costs. A strategy that combines various methodological approaches, including high-level conventional methods, may be advisable for assessing overall accuracy, particularly when a new class of problems is first addressed.

Acknowlgements. We are grateful to S. Krüger and S. Nagel for assistance during the preparation of the manuscript. A.G. thanks the Deutsche Forschungsgemeinschaft for a Heisenberg fellowship. This work was supported in part by the Deutsche Forschungsgemeinschaft, the German Bundesministerium für Bildung, Wissenschaft, Forschung und Technologie (grant no. 03D0050B), and the Fonds der Chemischen Industrie.

References

1. Hoffmann R (1982) Angew Chem Int Ed Engl 21:711
2. Albright TA, Burdett JK, Whangbo MH (1985) Orbital interaction in chemistry. Wiley, New York
3. Hofmann P (1986) Applied MO theory: Organometallic structure and reactivity problems. In: Veillard A (ed) Quantum chemistry: The challenge of transition metals and coordination chemistry. NATO Advanced Science Institute Series C176. Reidel, Dordrecht, p 253
4. Hay PJ, Wadt WR (1985) J Chem Phys 82:271
5. Frenking G, Antes I, Böhme M, Dapprich S, Ehlers AW, Jonas V, Neuhaus A, Otto M, Stegmann R, Veldkamp A, Vyboishchikov SF (1996) Pseudopotential calculations of transition metal compounds: Scope and limitations. In: Lipkowitz KB, Boyd DB (eds) Reviews in computational chemistry, Vol 8. VCH, New York, p 63
6. Almlöf J, Faegri Jr K, Korsell K (1982) J Comput Chem 3:385
7. Häser M, Ahlrichs R J Comput Chem (1989) 10:104
8. Siegbahn PEM (1983) The direct CI method. In: Diercksen GHF, Wilson S (eds) Methods in computational molecular physics. NATO Advanced Science Institute Series C113. Reidel, Dordrecht, p 189
9. Siegbahn PEM (1992) The configuration interaction method. In: Roos BO (ed) Lecture notes in quantum chemistry. Springer, Berlin Heidelberg New York, p 255
10. (a) Gregory K, Schleyer PvR, Snaith R (1991) Adv Inorg Chem 37:47; (b) Kaufmann E, Raghavachari K, Reed AE, Schleyer PvR. (1988) Organometallics 7:1597
11. Ziegler T (1991) Chem Rev 91:651
12. Veillard A (ed) (1986) Quantum chemistry: The challenge of transition metals and coordination chemistry. NATO Advanced Science Institute Series C176. Reidel, Dordrecht
13. Rösch N, Jörg H, Dunlap BI (1986) Application of the LCGTO-Xα method to transition metal carbonyls. In: Veillard A (ed) Quantum chemistry: The challenge of transition metals and coordination chemistry. NATO Advanced Science Institute Series C176. Reidel, Dordrecht, p 179
14. Baerends EJ, Rozendaal A (1986) Analysis of σ-bonding, π-(back) bonding and the synergetic effect in $Cr(CO)_6$. Comparison of Hartree-Fock and Xα results for metal-CO bonding. In: Ref. Veillard A (ed) Quantum chemistry: The challenge of transition metals and coordination chemistry. NATO Advanced Science Institute Series C176. Reidel, Dordrecht, p 159
15. Rösch N, Jörg H (1986) J Chem Phys 84:5967
16. Salahub DR, Zerner MC (eds) (1989) The challenge of d and f electrons. Theory and computation. ACS Symposium Series 394. American Chemical Society, Washington, DC

17. Labanowski JK, Andzelm JW (eds) (1991) Density functional methods in chemistry. Springer, Berlin Heidelberg New York
18. Verluis L, Ziegler T (1988) J Chem Phys 88:322
19. Fournier R, Andzelm J, Salahub DR (1989) J Chem Phys 90:6371
20. Dunlap BI, Andzelm J, Mintmire JW (1990) Phys Rev A 42:6354
21. Johnson BG, Gill PMW, Pople JA (1993) J Chem Phys 98:5612
22. Becke AD (1993) J Chem Phys 98:1372
23. Becke AD (1993) J Chem Phys 98:5648
24. Slater JC (1968) Quantum theory of matter, 2nd ed. McGraw-Hill, New York
25. Szabo A, Ostlund NS (1989) Modern quantum chemistry. McGraw Hill, New York
26. Parr RG, Yang W (1989) Density-functional theory. Oxford University Press, Oxford
27. Jones RO, Gunnarson O (1989) Rev Mod Phys 61:689
28. Dreizler RM, Gross EKU (1990) Density functional theory. Springer, Berlin Heidelberg New York
29. Trickey SB (ed) (1990) Density functional theory of many-fermion systems. Adv Quantum Chem 21. Academic Press, San Diego
30. Gross EKU, Dreizler RM (eds) (1995) Density functional theory. NATO Advanced Science Institute Series B337. Plenum Press, New York
31. Seminario JM, Politzer P (eds) (1995) Modern density functional theory. Elsevier Science, Amsterdam
32. Seminario JM (ed) (1996) Recent development and aplications of modern density functional theory. Elsevier Science, Amsterdam
33. Kohn W, Becke AD, Parr RG (1996) J Phys Chem 100:12974
34. Bartolotti LJ, Flurchick K (1996) Introduction to density functional theory. In: Lipkowitz KB, Boyd DB (eds) Reviews in computational chemistry, vol 7. VCH, New York, p 187
35. Curtiss LA, Raghavachari K, Trucks GW, Pople JA (1991) J Chem Phys 94:7221
36. Löwdin PO (1955) Phys Rev 97:1474
37. Bartlett RJ (1981) Ann Rev Phys Chem 32:359
38. Bartlett RJ, Stanton JF (1994) Applications of post-Hartree-Fock methods: A tutorial. In: Lipkowitz KB, Boyd DB (eds) Reviews in computational chemistry, vol 5. VCH Publishers, New York, p 65
39. Werner HJ (1987) Matrix-formulated direct multiconfiguration self-consistent field and multiconfiguration reference configuration-interaction methods. In: Lawley KP (ed) Ab initio methods in quantum chemistry, part II, Adv Chem Phys 69:1
40. Roos BO (1987) The complete active space self-consistent field method and its applications in electronic calculations. In: Lawley KP (ed) Ab initio methods in quantum chemistry, part II, Adv Chem Phys 69:399
41. Siegbahn PEM (1996) Electronic structure calculations for molecules containing transition metals. In: Prigogine I, Rice SA (eds) New methods in computational quantum mechanics. Adv Chem Phys 93:333
42. Kohn W, Sham LJ (1965) Phys Rev 140:A1133
43. Hohenberg P, Kohn W (1964) Phys Rev 136:B864
44. Lieb EH (1981) Rev Mod Phys 53:603
45. van Wüllen C, private communication
46. Levy M (1979) Proc Nat Acad Sci 76:6002
47. Levy M (1990) Constrained-search formulation and recent coordinate scaling in density-functional theory. In: Trickey SB (ed) Density functional theory of many-fermion systems. Adv Quantum Chem 21. Academic Press, San Diego, p 69
48. Dirac PAM (1930) Proc Cambridge Phil Soc 26:376
49. Jones RS, Trickey SB (1987) Phys Rev B 36:3095
50. Burke K, Perdew JP, Levy M (1995) Semilocal density functionals for exchange and correlation: Theory and applications. In: Seminario JM, Politzer P (eds) Modern density functional theory. Elsevier Science, Amsterdam, p 29

51. Talman JD, Shadwick WF, (1976) Phys. Rev. A 14:36
52. Kotani T (1995) Phys Rev Lett 74:2989
53. Städele M, Majewski JA, Vogl P, Görling A (1997) Phys Rev Lett 79:2089
54. Krieger JB, Li Y, Iafrate GJ (1992) Phys Rev A 45:101
55. Grabo T, Gross EKU (1997) Int J Quantum Chem 64:95
56. Slater JC, Wood JH (1971) Int J Quantum Chem Symp 4:3
57. Janak JF (1978) Phys. Rev. B 18:7165
58. Schlüter M, Sham LJ (1990) Density functional theory of the band gap. In: Trickey SB (ed) Density functional theory of many-fermion systems. Adv Quantum Chem 21. Academic Press, San Diego, p 97
59. Perdew JP (1990) Size-consistency, self-interaction correction, and derivative discontinuity in density functional theory. In: Trickey SB (ed) Density functional theory of many-fermion systems. Adv Quantum Chem 21. Academic Press, San Diego, p 113
60. Görling A (1996) Phys Rev A 54:3912
61. Harris J, Jones RO (1974) J Phys F: Solid State Phys 4:1170
62. Langreth DC, Perdew JP (1975) Solid State Commun 17:1425
63. Gunnarson O, Lundqvist (1976) Phys Rev B 13:4274
64. (a) Savin A, Umrigar CJ, Gonze X (1998) Chem Phys Lett 288:391; (b) Umrigar CJ, Savin A, Gonze X (1998) Are unoccupied Kohn-Sham eigenvalues related to excitation energies? In: Dobson JF, Vignale G, Das MP (eds) Electronic density functional theory: recent progress and new directions. Plenum, New York, p 167
65. Filippi C, Umrigar CJ, Gonze X (1997) J Chem Phys 107:9994
66. See, for example, Salem L (1966) Molecular orbital theory of conjugated systems, Benjamin, New York, p 110
67. Hoffmann R (1963) J Chem Phys 39:1397
68. Mulliken RS (1952) J Am Chem Soc 64:811
69. Pearson RG (1963) J Am Chem Soc 85:3533
70. Pearson RG (1997) Chemical hardness. Applications from molecules to solids. Wiley-VCH, Weinheim
71. Parr RG, Yang W (1989) Density-functional theory. Oxford University Press, Oxford, chaps 4 and 5
72. Geerlings P, De Proft F, Martin JML (1996) Density-functional theory concepts and techniques for studying molecular charge distributions and related properties. In: Seminario JM (ed) Recent development and aplications of modern density functional theory. Elsevier Science, Amsterdam, p 773
73. Nalewajski RF (1990) Charge sensitivity analysis as diagnostic tool for predicting trends in chemical reactivity. In: Ref [31] p 339
74. Slater JC (1974) The self-consistent field for molecules and solids: Quantum theory of molecules and solids, Vol IV. McGraw-Hill, New York
75. Savin A (1995) Beyond the Kohn-Sham determinant. In: Chong DP (ed) Recent advances in density functional methods. World Scientific, Singapore, p 129
76. Ziegler T, Rauk A, Baerends EJ (1977) Theor Chim Acta (Berl) 43:261
77. Barth U von (1979) Phys Rev A 20:1693
78. Filatov M, Shaik S (1998) Chem Phys Lett 288: 689
79. Stückl AC, Daul CA, Güdel HU (1997) Int J Quantum Chem 61:579 and references cited therein
80. Weiner B, Trickey SB (1998) Int J Quantum Chem 69:451
81. Görling A (1993) Phys Rev A 47:2783
82. Dunlap BI (1991) Symmetry and local potential methods. In: Labanowski JK, Andzelm JW (eds) Density functional methods in chemistry. Springer, Berlin Heidelberg New York, p 49
83. Löwdin PO (1963) Rev Mod Phys 35:496
84. Bauernschmitt R, Ahlrichs R (1996) J Chem Phys 104:9047
85. Becke AD (1988) Phys Rev A 38:3098

86. van Leeuwen R, Baerends EJ (1994) Phys Rev A 49:2421
87. Ernzerhof M, Burke K, Perdew JP (1996) Density functional theory, the exchange hole, and the molecular bond. In: Seminario JM (ed) Recent development and aplications of modern density functional theory. Elsevier Science, Amsterdam, p 207
88. Umrigar CJ, Gonze X (1994) Phys Rev A 50:3827
89. Vosko SH, Wilk L, Nusair M (1980) Can J Phys 58:1200
90. Perdew JP, Wang Y (1992) Phys Rev B 45:13244
91. Perdew JP, Zunger A (1981) Phys Rev B 23:5048
92. Kristyan S, Pulay P (1994) Chem Phys Lett 229:175
93. Pérez-Jordá JM, Becke AD (1995) Chem Phys Lett 233:134
94. Hedin L, Lundqvist BI (1971) J Phys C: Solid State Phys 4:2064
95. Moruzzi VL, Janak JF, Williams AR (1978) Calculated electronic properties of metals. Pergamon, New York, p 26
96. Perdew JP (1991) Unified theory of exchange and correlation beyond the local density approximation. In: Ziesche P, Eschrig H (eds) Electronic structure of solids. Akademie Verlag, Berlin
97. Perdew JP, Chevary JA, Vosko SH, Jackson KA, Pederson MR, Singh DJ, Filohais C (1992) Phys Rev B 46:6671
98. Perdew JP, Burke K, Ernzerhof M (1996) Phys Rev Lett 77:3865; (1997) Phys Rev Lett 78:1396 (E)
99. Perdew JP, Wang Y (1986) Phys Rev B 33:8800; (1989) Phys Rev B 40:3399
100. Perdew JP (1986) Phys Rev B 33:8822
101. Zhang Y, Yang W (1998) Phys Rev Lett 80:890
102. Perdew JP, Burke K, Ernzerhof M (1998) Phys Rev Lett 80:891
103. Lee C, Yang W, Parr RG (1988) Phys Rev B 37:785
104. Colle R, Salvetti O (1975) Theor Chim Acta 37:329
105. Morrison RC (1993) Int J Quantum Chem 46:583
106. Proynov EI, Salahub DR (1994) Phys Rev B 49:7874
107. Proynov EI, Vela A, Salahub DR (1995) Chem Phys Lett 230:419; Erratum (1995) ibid. 234:462
108. Proynov EI, Ruiz E, Vela A, Salahub DR (1995) Internat J Quantum Chem S29:61
109. Proynov EI, Sirois S, Salahub DR (1996) Internat J Quantum Chem 64:427
110. Filippi C, Gonze X, Umrigar CJ (1996) Generalized gradient approximations to density functional theory: comparison with exact results. In:Seminario JM (ed) Recent development and aplications of modern density functional theory. Elsevier Science, Amsterdam, p 295
111. Towler MD, Zupan A, Causa M (1996) Comp Phys Comm 98:181. Appendix A
112. Barone V (1995) Structure, magnetic properties, and reactivities of open-shell species from density functional and self-consistent hybrid methods. In: Chong DP (ed) Recent advances in density functional methods. World Scientific, Singapore, p 287
113. Seminario JM, (1990) An introduction to density functional theory in chemistry. In: Ref [31] p 1
114. Frisch MJ, Trucks GW, Schlegel HB, Gill PMW, Johnson BG, Robb MA, Cheeseman JR, Keith T, Petersson GA, Montgomery JA, Raghavachari K, Al-Laham MA, Zakrzewski VG, Ortiz JV, Foresman JB, Cioslowski J, Stefanov BB, Nanayakkara A, Challacombe M, Peng CY, Ayala PY, Chen W, Wong MW, Andres JL, Replogle ES, Gomperts R, Martin RL, Fox DJ, Binkley JS, Defrees DJ, Baker J, Stewart JP, Head-Gordon M, Gonzalez C, Pople JA (1995) Gaussian 94, Gaussian Inc, Pittsburgh PA
115. Görling A, Ernzerhof M (1995) Phys Rev A 51:4501
116. Görling A, Levy M (1997) J Chem Phys 106:2675
117. Pople JA, Head-Gordon M, Raghavachari K (1987) J Chem Phys 87:5968
118. (a) Curtiss LA, Raghavachari K, Pople JA (1995) J Chem Phys 103:4192; (b) Raghavachari K, Curtiss LA (1995) Evaluation of bond energies to chemical accuracy by quan-

tum chemical. In: Yarkony DR (ed) Modern electronic structure theory. World Scientific, Singapore, p 991

119. (a) Siegbahn PEM, Blomberg MRA, Svensson M (1994) Chem Phys Lett 223:35; (b) Siegbahn PEM, Svensson M, Boussard PJE (1995) J Chem Phys 102:5377
120. Ceperley DM, Alder BJ (1980) Phys Rev Lett 45:566
121. Becke AD (1992) J Chem Phys 96:2155
122. (a) Peterson GA, Al-Laham MA (1991) J Chem Phys 94:6081; (b) Peterson GA, Bennett A, Tensfeldt TG, Al-Laham MA, Shirley WA, Mantzaris J (1988) J Chem Phys 89:2193
123. Curtiss LA, Raghavachari K, Redfern PC, Pople JA (1997) J Chem Phys 106:1063
124. Baker J, Muir M, Andzelm J (1995) J Chem Phys 102:2063
125. Baker J, Muir M, Andzelm J, Scheiner A (1996) Hybrid Hartree-Fock density-functional theory functionals: the adiabatic connection method. In: Laird BB, Ross RB, Ziegler T (eds) Chemical applications of density-functional theory. ACS Symposium Series 629, American Chemical Society, Washington DC, p 342
126. Stanton RV, Merz Jr KM (1994) J Chem Phys 100:434
127. Jursic BS (1996) Computing transition state structure with density functional theory methods. In: Seminario JM (ed) Recent development and aplications of modern density functional theory. Elsevier Science, Amsterdam, p 709
128. Johnson BG, Gonzales CA, Gill PMW, Pople JA (1994) Chem Phys Lett 221:100
129. Gunnarsson O, Jones RO (1995) Phys Rev B 31:7588
130. Ziegler T, Li J (1994) Can J Chem 72:783
131. Blomberg MRA, Siegbahn PEM, Svensson M (1996) J Chem Phys 104:9546
132. Rösch N, Trickey SB (1997) J Chem Phys 106:8940
133. (a) Ehlers A, Frenking G (1993) J Chem Soc Chem Commun: 1709; (b) Ehlers A, Frenking G (1994) J Am Chem Soc 116:1514
134. (a) Blomberg MRA, Brandemark UB, Siegbahn PEM, Wennerberg J, Bauschlicher Jr CW (1988) J Am Chem Soc 110:6650; (b) Bauschlicher Jr CW, Langhof SR (1989) Chem Phys 129:431
135. Barnes LA, Rosi M, Bauschlicher Jr CW (1991) J Chem Phys 94:2031
136. Blomberg MRA, Siegbahn PEM, Lee TJ, Rendell AP, Rice JE (1991) J Chem Phys 95:5898
137. (a) Stevens AE, Feigerle CS, Lineberger WC (1982) J Am Chem Soc 104:5026; (b) Lewis KE, Golden DM, Smith GP (1984) J Am Chem Soc 106:3905
138. Li J, Schreckenbach G, Ziegler T (1995) J Am Chem Soc 117:486
139. Jonas V, Thiel W (1995) J Chem Phys 102:8474
140. Fournier R, Pápai I (1995) Infrared spectra and binding energies of transition metal-monoligand complexes. In: Chong DP (ed) Recent advances in density functional methods. World Scientific, Singapore, p 219
141. Nasluzov VA, Rösch N (1996) Chem Phys 210:413
142. Bauschlicher Jr CW, Partridge H, Sheehy JA, Langhoff SR, Rosi M (1992) J Phys Chem 96:6969
143. Holthausen MC, Heinemann C, Cornehl HH, Koch W, Schwarz H (1995) J Chem Phys 102:4931
144. Holthausen MC, Mohr M, Koch W (1995) Chem Phys Lett 240:245
145. Ricca A, Bauschlicher CW Jr (1995) Chem Phys Lett 245:150
146. Rohlfing CM, Hay PJ (1985) J Chem Phys 83:4641
147. Lüthi HP, Siegbahn PEM, Almlöf J (1985) J Phys Chem 89:2156
148. Barnes LA, Liu B, Lindh R (1993) J Chem Phys 98:3978
149. Jonas V, Thiel W (1996) J Chem Phys 105:3636
150. Bray MR, Deeth RJ, Paget VJ, Sheen PD (1996) Int J Quantum Chem 61:85
151. Bérces A, Ziegler T (1994) J Phys Chem 98:13233
152. Bérces A (1996) J Phys Chem 100:16538
153. Wachters AJH (1970) J Chem Phys 52:1033
154. Dunning Jr TH (1971) J Chem Phys 55:716

155. (a) Dolg M, Wedig U, Stoll H, Preuß H (1987) J Chem Phys 86:866; (b) Andrae D, Häußermann U, Dolg M, Stoll H, Preuß H (1990) Theor Chim Acta 77:123
156. (a) Delley B, Wrinn M, Lüthi HP (1994) J Chem Phys 100:5785; (b) Delley B (1994) DMol a Standard tool for density functional calculations: review and advances. In: Seminario JM, Politzer P (eds) Density functional theory. Elsevier Science, Amsterdam p 221
157. (a) Braterman PS (1975) Metal carbonyl spectra. Academic, London; (b) Jones LH, McDowell RS, Goldblatt M (1969) Inorg Chem 8:2349
158. Merle-Mejean T, Cosse-Mertens C, Bouchareb S, Galan F, Mascetti J, Tranquille M (1992) J Phys Chem 96:9148
159. Rösch N, Krüger S, Mayer M, Nasluzov VA (1996) The Douglas-Kroll-Hess approach to relativistic density functional theory: methodological aspects and applications to metal complexes and clusters. In: Seminario JM (ed) Recent development and aplications of modern density functional theory. Elsevier Science, Amsterdam, p 497
160. Dunning Jr TH, Hay PJ (1976) Gaussian basis sets for molecular calculations. In: Schaefer III HF (ed) Modern theoretical chemistry, Vol 3, New York, p 1; (b) Hay PJ, Wadt WR (1985) J Chem Phys 82:299
161. (a) Kolb HC, VanNieuwenzhe MS, Sharpless KB (1994) Chem Rev 94:2483 (b) Johnson RA, Sharpless KB (1993) Catalytic asymmetric dihydroxylation. In:. Ojima I (ed) Catalytic asymmetric synthesis. VCH, New York
162. (a) Jacobsen EN (1993) Asymmetric catalytic epoxidation of unfunctionalized olefins. In: Ojima I (ed) Catalytic asymmetric synthesis. VCH, New York; (b) Sheldon RA (1992) Catalytic oxidations with hydrogen peroxides as oxidant. Kluwer, Rotterdam
163. Jorgensen KA (1989) Chem Rev 89:431
164. (a) Herrmann WA (1995) J Organomet Chem 500:149; (b) Herrmann WA, Kühn FE (1997) Acc Chem Res 30:169; (c) Romão CC, Kühn FE, Herrmann WA (1997) Chem Rev 97:3197
165. Pidun U, Boehme C, Frenking G (1996) Angew Chem Int Ed Engl 35:2817
166. Dapprich S, Ujaque G, Maseras F, Lledos A, Musaev DG, Morokuma K (1996) J Am Chem Soc 118:11660
167. Torrent M, Deng L, Duran M, Sola M, Ziegler T (1997) Organometallics 16:13
168. DelMonte AJ, Haller J, Houk KN, Sharpless KB, Singleton DA, Strassner T, Thomas AA (1997) J Am Chem Soc 119:9907
169. Gisdakis P, Antonczak S, Köstlmeier S, Herrmann WA, Rösch N (1998) Angew Chem 110:2331
170. Wu YD, Sun J (1998) J Org Chem 63:1752
171. Criegee R, Marchand B, Wannowius H (1942) Justus Liebigs Ann Chem 550:99
172. Herrmann WA, Fischer RW, Scherer W, Rauch MU (1993) Angew Chem Int Ed Engl 32:1157
173. (a) Al-Ajlouni AM, Espenson JH (1995) J Am Chem Soc 117:9243; (b) Al-Ajlouni AM, Espenson JH (1996) J Org Chem 61:3969
174. Köstlmeier S, Nasluzov VA, Herrmann WA, Rösch N (1997) Organometallics 16:1786
175. Sharpless KB, Townsend JM, Williams DR (1972) J Am Chem Soc 94:195
176. Mimoun H, de Roch IS, Sajus L (1970) Tetrahedron 26:37
177. Thiel WR (1996) Chem Ber 129:575
178. Gisdakis P, Rösch N unpublished results
179. (a) Singleton DA, Merrigan SR, Liu J, Houk KN (1997) J Am Chem Soc 119:3385; (b) Bach RD, Estévez CM, Winter JE, Glukhovtsev MN (1998) J Am Chem Soc 120:680

Hybrid Quantum Mechanics/Molecular Mechanics Methods in Transition Metal Chemistry

Feliu Maseras
e-mail: feliu@klingon.uab.es
Unitat de Química Física, Edifici C.n, Universitat Autònoma de Barcelona,
E-08193 Bellaterra, Catalonia, Spain

Hybrid quantum mechanics/molecular mechanics (QM/MM) methods are making a powerful entry in the field of computational organometallic chemistry. This chapter reviews the current status of the field. The methodological features more relevant to the organometallic chemist are briefly discussed, with special emphasis on the significance of geometry optimization within the hybrid QM/MM scheme. The peculiarities of the Integrated Molecular Orbital Molecular Mechanics (IMOMM) method, which is one of the most commonly applied, are presented in some detail. The published applications of hybrid QM/MM methods to transition metal chemistry are also reviewed. The presentation is classified by chemical topic, and results on structural problems, olefin polymerization, asymmetric dihydroxylation, agostic complexes and bioinorganic complexes are reviewed. The method is shown to be especially efficient in the quantitative introduction of steric effects. Its capabilities for the analysis of results, both in the quantitative separation of electronic and steric effects and in the identification of the atoms involved in steric effects, are emphasized.

Keywords: Hybrid QM/MM methods, IMOMM method, Steric effects, Quantum chemistry, Molecular mechanics

Topics in Organometallic Chemistry, Vol. 4
Volume Editors: J.M Brown and P. Hofmann
© Springer-Verlag Berlin Heidelberg 1999

1
Introduction

There is nowadays a wide and expanding variety of methods in computational chemistry, ranging from the very approximate to the very precise. Computer power has been growing steadily through the years, and it seems that it will not stop increasing in the foreseeable future. So an external observer could get the impression that performance of computations on any system at the highest computational level will be possible in a short time, and that one should not worry any more about the design of approximate methods. This situation is however far from reality.

A large portion of the computational methods, and certainly the most accurate, rely on quantum mechanics (QM). They allow an accurate description of the electronic structure of molecules. The more precise methods are only feasible in practice for small molecules, its computational cost increasing sharply with the size of the system. So the computational chemist is usually confronted with the problem of choosing the method able to provide a sufficient quality for each chemical question. This problem is especially severe in the field of transition metal complexes. The usual presence of large ligands rapidly increases the size of the systems, and the electronic complexity of transition metal atoms requires the use of high accuracy methods.

Throughout the years, the calculation of transition metal complexes has been treated from a number of points of view. The early stages were marked by the application of the Extended Hückel theory [1], which led to a number of interesting results, though always of a qualitative nature [2]. Later on, ab initio quantum chemistry methods have been applied, either based on the Hartree-Fock (HF) theory [3] or on the density functional theory (DFT) [4]. Nowadays, it is generally accepted that the minimal approaches required in order to get reliable results are the introduction of dynamic correlation (in the HF theory) and the use of non-local functionals (in the DFT theory). The performance of QM methods in organometallic systems is reviewed in other chapters of this volume. Even if these methods are usually an optimal choice in terms of quality, they are certainly demanding in terms of computer time, and often become prohibitive when bulky ligands are present, which is often the case in organometallic chemistry.

An approach that has been extensively applied to circumvent this problem is the use of model systems [5, 6]. The calculation is not carried out on the real system but on a model system that has hopefully the same behavior. These model systems are obtained through simplification of the ligands of the system, usually in the regions further away from the metal center. For instance, the researcher interested in the $Pt(P(t\text{-}Bu)_3)_2$ system would compute instead the $Pt(PH_3)_2$ system, where the *tert*-butyl groups attached to phosphorus are replaced by hydrogen atoms. This radically simplified approach is nevertheless very efficient in a number of cases. The reason for this efficiency is that the metal-ligand interaction remains well reproduced. In the example mentioned above, the $Pt\text{-}P(t\text{-}Bu)_3$ bond is not very different from the Pt-PH$_3$ bond, and because of that the electronic properties of the metal atom remain mostly unchanged. This approach is obviously not so good as far as the ligand-ligand interactions are concerned. Therefore, the challenge of studying problems where ligand-ligand interactions, essentially of steric nature, play an important role cannot be met by calculations on model systems.

If the system cannot be reduced in size, and it is too large to be computed through an ab initio QM calculation, other possibilities have to be explored. One possible venue within the domain of QM calculations is the use of semiempirical methods specifically tailored for transition metal systems. Several attempts have been made in this direction, with promising results from the PM3(tm) method [7], but more research is still required in this area.

Another major venue is the use of molecular mechanics (MM) methods [8–10]. Molecular mechanics is a simple, empirical "ball-and-spring" model of molecular structure. Atoms (balls) are connected by springs (bonds) that can be stretched or compressed by intramolecular forces. The size of the balls and the rigidity of the springs are determined empirically, that is, they are chosen to reproduce experimental data. Electrons are not part of the MM description, and as a result this technique is unable to describe several fundamental phenomena, like electronic spectroscopy of photochemistry. Nevertheless, MM methods are orders of magnitude cheaper than QM methods, and as such they have found a prominent place in several domains of computational chemistry, like conformational search of organic compounds or molecular dynamics [11–13]. One of its strong points is actually the quantitative introduction of steric effects.

Application of pure MM methods to transition metal systems is hampered by the relative scarceness of experimental data. There is a much larger variety of elements and binding modes than in organic chemistry, and much fewer examples of each class of compound. As a result, parameterization is much more complicated, and the validity of force fields is more limited than that of the all-purpose force fields used for organic compounds. One possible solution is the use of QM calculations on model systems to adjust the MM parameters involving the transition metal atom [14, 15]. This approach can yield precise MM parameters, which can be used subsequently for extremely fast calculations. The downside is that the parameterization process is very computer demanding, much more than that of a normal calculation on the model system, and that these parameters are valid for a very limited range of systems.

A logical answer to all the problems described above for the calculation of large transition metal systems is the use of QM methods for the part of the molecule involving the metal, and of MM methods for the part removed from the metal. This is precisely the general idea of hybrid QM/MM methods. Following the same example mentioned above, the hybrid QM/MM calculation on the Pt(t-Bu$_3$)$_2$ system would use a QM level to describe the interaction between platinum and the phosphines, and an MM level to describe the steric interaction between the bulky t-Bu substituents of the phosphine ligands.

Hybrid QM/MM methods already have a certain history of their own in computational chemistry [16–21]. Most work on this field has been on the introduction of solvation effects, with special focus on biochemical systems. A usual partition is to have the solvate computed at the QM level and the solvent molecules introduced at the MM level. If the solvate is composed exclusively of a few organic molecules, the QM part of the calculation can indeed be carried out very efficiently through the use semiempirical methods. As a result, hybrid QM/MM calculations have been used extensively in the performance of molecular dynamics calculations of solvation effects, and there are indeed a variety of different methods to carry out this type of calculation.

The application of the QM/MM algorithms traditionally applied in the description of solvation to transition metal chemistry so far has not been very productive. There is a major difference concerning the nature of the interaction between the QM and MM parts of the systems. While in a solvate/solvent system there are no bonds between the two parts, in a typical transition metal complex there are chemical bonds crossing the interface between the QM and MM regions of the molecule. The design and application of QM/MM methods in systems where there are chemical bonds between the two different regions of the molecule is more recent. Two major lines of applications seem to be developing, one concerning zeolite solid systems [22–24] and another concerning transition metal complexes [25–27]. This chapter will try to review the current state of the second of these fields. After this introduction, the second section will deal briefly with the technical aspects of the different methods, the third section will present an overview of the published applications, and the fourth section will bring together the conclusions and future perspectives.

2
Hybrid QM/MM Methods

The simple idea of partitioning a chemical system in QM and MM regions gives rise to a number of different methodologies depending on the particular way the calculation is performed. It is not the goal of this chapter to enter into the technical details in depth, but a minimum description is necessary for the reader to evaluate the quality of different calculations that can be found in the chemical literature.

2.1
General Overview

A very minimal mathematical background is given in this subsection to allow the understanding of the rest of the section. More detailed presentations can be found elsewhere [21, 22].

The common feature of all hybrid QM/MM methods is the partition of the system in at least two regions: one where the quantum mechanical description is applied (to be referred to as the QM region), and another one where the MM description is applied (to be referred to as the MM region). The calculation of the hybrid QM/MM energy of the whole system can be in all generality expressed as:

$$E_{tot}(QM,MM)=E_{QM}(QM)+E_{MM}(MM)+E_{interaction}(QM/MM)$$

where the subscript labels refer to the type of calculation and the labels in parenthesis correspond to the region under study. The interaction energy between the QM and MM regions has to be in principle computed by both the QM and MM methods, and the previous expression becomes:

$$E_{tot}(QM,MM)=E_{QM}(QM)+E_{MM}(MM)+E_{QM}(QM/MM)+E_{MM}(QM/MM)$$

The energy expression in a general hybrid QM/MM method thus has four components. Two of them are simply the pure QM and MM calculations of the corresponding regions, and their computation is straightforward. The other two correspond to the evaluation of the interaction between both regions, in principle at both computational levels. Different computational schemes are defined by the choice of method to compute the $E_{QM}(QM/MM)$ and $E_{MM}(QM/MM)$ terms.

The $E_{QM}(QM/MM)$ term accounts for the direct effect of the atoms in the MM region on the QM energy of the system. This term is usually critical in solvation problems, because one of the points of interest is precisely how the wavefunction of the solute is affected by the presence of the solvent. In the case of a transition metal complex, this term accounts mainly for the electronic effects of the ligand substituents on the metal center.

A simple way to introduce the $E_{QM}(QM/MM)$ term is to put electrostatic charges in the positions occupied by the MM atoms [21]. One initial problem with this kind of approach is the choice of the electrostatic charges, which is by no means trivial. A more serious problem appears when there are chemical bonds between the QM and MM regions, where this approach breaks down in the proximity of the interface and needs to be reformulated. A more elaborate scheme that overcomes this limitation has been proposed through the introduction of localized orbitals [28–31], although it has been applied so far mostly on organic systems. The Integrated Molecular Orbitals Molecular Mechanics (IMOMM) scheme [25], which will be discussed below in detail, simply neglects this $E_{QM}(QM/MM)$ term.

The $E_{MM}(QM/MM)$ term accounts for the direct effect of the atoms in the QM region in the MM energy of the system. This term is usually important in transition metal complexes, because it accounts for the steric constraints introduced by the presence of a metal center on the geometry of the ligands. In other words, it is mostly related to the steric effects of the ligand substituents. The $E_{MM}(QM/MM)$ term is usually introduced through the parameterization of the QM atoms with the same force field used in the MM region. When there are no bonds between the QM and MM regions, it corresponds simply to non-bonded interactions between the QM and MM regions, with things being slightly more complicated when the two regions are connected through chemical bonds.

It has been repeatedly mentioned in the previous paragraphs that the presence of chemical bonds between the QM and MM regions poses a problem with the use of hybrid QM/MM methods. One of the reasons is that most approaches require the introduction of additional atoms (usually hydrogen) in the QM calculation because of the impossibility of having dangling bonds. The way to deal with these additional atoms introduces another wide range of methodological details that will not be discussed here.

2.2
One-Step QM/MM Methods: Combined MO+MM Method

The simplest approach to the use of hybrid QM/MM methods is the performance of two independent calculations. First, the geometry of the atoms in the QM region is optimized in a pure QM calculation. Afterwards, the real ligands are added in an MM calculation where the structure of the atoms in the model system is kept frozen. This approach has its big advantage in the simplicity: the researcher needs only access to two independent standard programs able to carry out the QM and the MM calculations. No special computer code is required, and the addition of the results of both calculations can be done by hand.

In the terms described in the previous subsection, this approach has a series of implications. The first is the neglect of the $E_{QM}(QM/MM)$ term. This is usually reasonable in transition metal complexes, where the MM regions has mostly a steric effect. A more serious limitation of this approach is the lack of relaxation of atoms in the QM region. They stay frozen at the geometry of the model system, regardless of any steric strain that might exist on them. These methods are labeled here as one-step methods to distinguish them from multistep methods, where the geometry of the QM part is relaxed through an iterative procedure. They have also been labeled elsewhere as combined MO+MM methods [32–34] and MO-then-MM methods [26].

Houk and coworkers in the 1980s had already applied one-step hybrid QM/MM methods to the study of organic systems [35, 36]. Later on, they were applied to transition metal systems in a number of cases. In particular, one-step QM/MM methods have been used extensively by Morokuma and coworkers in the study of a number of problems, like olefin polymerization [33, 37, 38], and fluxionality of hydride complexes of iron [32]. Other studies with this type of method were carried out by Eisenstein and coworkers on the structure of $OsCl_2H_2(P(i\text{-}Pr)_3)_2$ systems [39] and by Maseras and Lledós on the fluxionality of $[Me_2Si(NSiMe_3)_2]_2InLi$ species [40]. The most significant of these calculations will be discussed in the next section together with other applications.

The particular features of one-step methods can be probably better understood with an example taken from a study by Matsubara et al. on the $Pt(PR_3)_2+H_2$ system (Scheme 1) [26]. In this study, which is in fact a calibration of the multistep IMOMM procedure, the authors compare the results obtained with different hybrid approaches with those of a much more expensive pure MO study of the complete system. In particular, they consider four different structures: the reactants, the *trans* product, the *cis* product, and the transition state connecting the reactants to the *cis* product. In the case of R=*t*-Bu, the relative energies at the HF level of each of these four stationary points are 0.0, -10.1, 18.5 and 33.6 kcal·mol^{-1}, respectively. The one-step hybrid QM/MM calculation carried out with a Pt(PH$_3$)+H$_2$ model for the QM part at the same computational level and the MM3 force field for the MM part yields values of 0.0, -13.9, 87.3 and 39.7 kcal·mol^{-1}. Relative energies computed with the hybrid method are thus quite close to those of the pure MO calculation (within 7 kcal·mol^{-1}) except for the case of the *cis* isomer, where the deviation is as large as 68.8 kcal·mol^{-1}. The reason for this huge difference can be easily traced back to the geometry distortion. The value computed at the pure MO level for the P-Pt-P angle is 104.7° in *cis*-Pt(PH$_3$)$_2$H$_2$, and 129.2° in the *cis*-Pt(P(t-Bu)$_3$)$_2$H$_2$ system. The one-step hybrid QM/MM calculation uses the value of the model system (104.7°) for the real system, resulting in a large steric repulsion that leads to the corresponding increase in the relative energy. Things are not so bad for the other stationary points, where the steric repulsion is smaller, and the P-Pt-P bond angles are more similar.

This example illustrates the advantages and limitations of one-step hybrid QM/MM methods. They are able to produce results in reasonable agreement with those of more elaborate methods and, as such, they cannot be neglected as a valuable tool. On the negative side, they can fail dramatically if there is an important distortion of the geometry of atoms in the QM region. An additional inconvenience of these methods is that no analysis can be carried out on the geometries of the QM atoms, simply because they are not optimized. A more troublesome problem is that the validity of the method cannot be assessed in the absence of external data, either coming from experiments or from more elaborate calculations. All these limitations are overcome by multistep QM/MM methods.

2.3
Multistep QM/MM Methods: The IMOMM Method

The natural extension to single-step methods is the use of multistep methods, where the geometry of the model system is corrected to account for the effect of atoms in the MM part. The basic equation of the energy in multistep methods is not very different from that of single-step methods, and it can indeed be the same if one neglects the E_{QM}(QM/MM) term, as is done in a number of schemes. However, there is a substantial difference in the choice of the geometries. In sin-

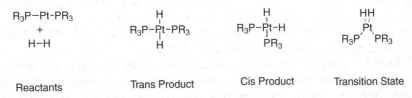

Scheme 1. The four structures considered in the study of the Pt(PR$_3$)$_2$+H$_2$ system [26]

gle-step methods the geometry of each region is optimized independently, in particular with the QM method for the QM region. As a result, in general, the final geometry is not the best possible one from the QM/MM point of view. In multistep methods the geometry optimization is complete within the QM/MM formalism, and as a result the best possible geometry is obtained and used for the computation of the energy. In other words, the fully optimized QM/MM geometry is neither the optimal QM nor the optimal MM arrangement of the atoms, but the arrangement that reaches the best compromise between the two of them.

The performance of a full geometry optimization has important implications both from a technical point of view and from the point of view of applications. From a technical point of view it requires the use of QM/MM gradients, the consistent computation of which from the pure QM and MM gradients has a number of pitfalls which will not be discussed in detail here. In any case, authors use different approaches to solve this problem, giving rise again to different methods.

The difference in the applications is illustrated again by the study of Matsubara et al. discussed above [26]. Let us recall that a one-step method yielded for the cis-Pt(P(t-Bu)$_3$)$_2$H$_2$ complex an energy 87.3 kcal·mol^{-1} above that of Pt(P(t-Bu)$_3$)$_2$+H$_2$, a result which was in sharp disagreement with the pure MO difference of 18.5 kcal·mol^{-1}. The use of a multistep hybrid QM/MM method is in very good agreement with the pure MO result, with the energy difference between both species falling to 17.2 kcal·mol^{-1}. This success is closely related to the improvement of the P-Pt-P angle by the multistep method, with a value of 124.9° (to be compared with 129.2° in the full MO calculation, and with the 104.7° used in the one-step calculation). Therefore, in this case, the geometry optimization feature of the multistep method is able to overcome the limitations of the one-step approach and give a correct result.

Within the variety of available multistep hybrid QM/MM methods [17, 18, 41, 42], the one that has been applied more often to transition metal complexes is the Integrated Molecular Orbitals Molecular Mechanics (IMOMM) method [25]. Its most notable characteristics are the following:

- It is a full multistep method. The final geometry is optimized within the hybrid QM/MM formalism.
- It is not associated with any particular QM method or force field. It can be used with any combination of them.
- It neglects the E_{QM}(QM/MM) term. Therefore, it does not introduce electronic effects of ligands. The effect of the ligands is nevertheless introduced indirectly in the QM energy through the geometry distortions they induce.
- In its original formulation it freezes the lengths of the chemical bonds between the QM and MM regions of the molecule.
- It includes an algorithm to perform a geometry optimization of the MM region (microiteration) in each step of the geometry optimization of the QM region (macroiteration). Although this technical feature greatly improves the computer efficiency, it does not alter the final outcome of the calculation.

The general philosophy of the IMOMM method is therefore to provide a simple and robust algorithm for the application of hybrid QM/MM methods to transition metal systems. This is achieved at the price of losing some of the precision, especially in what concerns the neglect of electronic effects and freezing of bonds between the QM and MM regions. A number of applications of the IMOMM method to transition metal chemistry will be presented in the next section. The QM and MM computational levels will be expressed in a compact way in the form IMOMM(QM-level:MM-level). Although they will not be discussed here, other applications of the IMOMM method have also been been carried out on organic systems [43–47].

2.4
Checklist of Technical Features

Since this chapter is intended for a general audience of organometallic chemists not specializing in computational chemistry, a brief summary of this section is given here in the form of a practical checklist of features indicative of the quality of a hybrid QM/MM calculation.

1. Partition. The reliability of the calculation depends in the first place on which atoms are in the QM part and which atoms in the MM part. The larger the QM calculation the better, because it is generally assumed that the QM calculation is more precise than the MM calculation.
2. Nature of the QM and MM methods. The label hybrid QM/MM encompasses all the possible quantum mechanics and force fields, with a large variety of qualities and computational prices. It is critical that both the QM and MM levels reproduce properly the main interactions in the respective regions. The use of a hybrid QM/MM method will not be able to overcome errors in the modeling of the different regions.
3. One-step or multistep method. One-step methods are simple and efficient in a number of cases, but they can give very poor results if geometry relaxation in the QM region is important. This problem is solved by multistep methods.
4. Introduction of electronic effects. If the E_{QM}(QM/MM) term is neglected, electronic effects are not introduced, and the calculation will consider only steric effects.

3
Applications

This section presents a summary of the applications that have been carried out on transition metal complexes. It is divided by chemical subject rather than by methodological type. It will be seen that although the initial calculations were carried out on problems where the presence of steric effects was apparent, the range of applications is rapidly expanding.

3.1
Structural Studies

A first set of applications of hybrid QM/MM methods to transition metal chemistry concerns the reproduction of geometrical distortions caused by steric effects. Geometry optimization is not a feature of one-step methods, requiring multistep methods. In particular, all applications discussed in this subsection are carried out with the IMOMM method. The objective of this type of study is twofold. On one hand, they constitute a validation of hybrid QM/MM methods through comparison of their results with those of crystal structures. On the other hand, in the particular case of IMOMM, they allow a quantitative separation of electronic and steric effects. IMOMM only introduces steric effects of ligands, and its success means therefore that electronic effects are of lesser importance.

Barea et al. [27] have analyzed steric effects on the $ReH_5(PPh(i\text{-}Pr)_2)_2(SiHPh_2)_2$ and $ReH_5(PCyp_3)_2(SiH_2Ph)_2$ systems (Fig. 1) at the IMOMM(MP2:MM3) computational level, the QM part being constituted by the $ReH_5(PH_3)_2(SiH_3)_2$ system. These complexes are experimentally known [48] to have a capped square antiprism structure, a quite common polyhedron for 9-coordinate species [49]. Despite the qualitative similarities, there are some quantitative differences between the crystal structures of both species, especially in what concerns the bond angles, the largest difference being in the Si-Re-Si angle, with a value of 117.9° for $ReH_5(PPh(i\text{-}Pr)_2)_2(SiHPh_2)_2$ and a value of 97.9° for $ReH_5(PCyp_3)_2(SiH_2Ph)_2$. This difference is reproduced by the IMOMM calculation, with computed values of 113.8° and 102.1°, respectively.

Apart from the reproduction of the geometrical features of these 9-coordinate species, the paper by Barea et al. [27] also exemplifies the possibilities of a quantitative analysis of steric effects implicit in a hybrid QM/MM calculation. In particular, in the case of $ReH_5(PCy_3)_2(SiH_2Ph)_2$, the leading steric interactions are

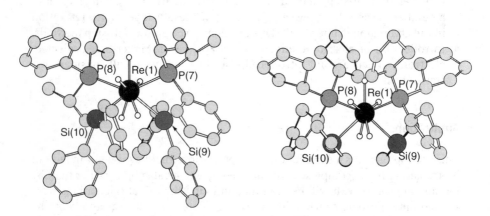

Fig. 1. IMOMM(MP2:MM3) optimized structure [27] of complexes $ReH_5(PPh(i\text{-}Pr)_2)_2$ $(SiHPh_2)_2$ (*left*) and $ReH_5(PCyp_3)_2(SiH_2Ph)_2$ (*right*). Hydrogen atoms not directly attached to the metal are omitted for clarity

shown to be repulsive, and to be associated with the P(7)-Si(9) and P(8)-Si(10) ligand pairs, which are related by symmetry, with values of 2.28 and 2.30 kcal·mol^{-1}. They are trailed a distance by the P(7)-P(7) and P(8)-P(8) pairs, corresponding to intraligand reorganization, with a weight of 0.61 kcal·mol^{-1}.

Apart from introducing this simple analysis technique, the paper by Barea et al. constitutes essentially a satisfactory calibration of the hybrid QM/MM method IMOMM against crystal structures on a system with small distortions. The test of its performance in cases with larger distortions can be found in a paper by Maseras and Eisenstein on the $OsCl_2H_2(P(i\text{-}Pr)_3)_2$ complex [50]. This 6-coordinate complex is known to present a trigonal prism crystal structure [51] (Fig. 2), quite unusual for a formally d^4 complex. A previous RHF calculation [39] on the $OsCl_2H_2(PH_3)_2$ system had yielded the more common bicapped tetrahedron geometry. The formal relationship between the bicapped tetrahedron and the trigonal prism is illustrated by the Cl4-Os-X-P2 dihedral angle (X=dummy atom on the bisector of the P2-Os-P3 angle). This angle has a value of 0° for the bicapped tetrahedron (Cl-Os-Cl plane parallel to the P-Os-P plane) and a value of 45° for the capped trigonal prism (Cl-Os-Cl plane halfway between the two perpendicular P-Os-P and H-Os-H planes).

The paper by Maseras and Eisenstein [50] presents full geometry optimizations at the Becke3LYP level for the $OsCl_2H_2(PH_3)_2$ system and at the IMOMM(Becke3LYP:MM3) level for the $OsCl_2H_2(P(i\text{-}Pr)_3)_2$ system. The Becke3LYP calculation on the model system yields the expected bicapped tetrahedron, with a Cl4-Os-X-P2 dihedral angle of 0.0°. This situation is substantially changed in the IMOMM(Becke3LYP:MM3) calculation on the real system. The complex is no longer a bicapped tetrahedron, and the Cl4-Os-X-P2 dihedral angle is 35.7°, which is much closer to the experimental value of 41.9°. This substantial improvement proves two different points. On one hand it is a satisfactory test for

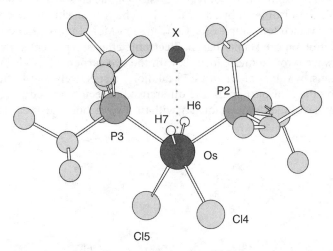

Fig. 2. IMOMM(Becke3LYP:MM3) optimized structure [50] of complex $OsCl_2H_2(P(i\text{-}Pr)_3)_2$. Hydrogen atoms not directly attached to the metal are omitted for clarity

the IMOMM method on this highly distorted system. On the other hand, it proves that the origin of the distortion is purely steric, since no electronic effects are introduced from the alkyl ligands.

The performance of the same quantitative analysis presented above for the 9-co-ordinate Re complexes produces interesting comparisons. Steric repulsion is more important in the Os complex, with the two largest terms corresponding to phos-phine-chloride repulsions: P3-Cl5 at 3.76 kcal·mol^{-1} and P2-Cl4 at 2.94 kcal·mol^{-1}. It is also worth noting the importance of the P2-P3 term (2.77 kcal·mol^{-1}), and that of the intraphosphine repulsions P2-P2 and P3-P3 (1.36 and 1.59 kcal·mol^{-1}). In summary, the existence of so many terms above 1 kcal·mol^{-1} is just a measure of the high steric strain in this complex.

The application of hybrid QM/MM methods to structural problems is not re-stricted to the reproduction of crystal structures. Ogasawara et al. [52] used IM-OMM(MP2:MM3) calculations to evaluate the relative energy of the two possible isomers of Ru(CO)$_2$(PR$_3$)$_3$ complexes. These d^8 ML$_5$ species have two main iso-meric forms A and B, both of them of trigonal bipyramidal nature (Scheme 2). The two forms can be observed experimentally [52, 53] by changing the nature of the phosphine ligand PR$_3$. In particular, when PR$_3$=PEt$_3$, the B isomer is the species present in the crystal, with a C-Ru-C bond angle of 133.6°. When PR$_3$= P(i-Pr)$_2$Me, two independent molecules are present in the crystal, one of them of A nature (C-Ru-C angle of 173.6°) and the other one of B nature (C-Ru-C angle of 146.7°). IMOMM(MP2:MM3) calculations were carried out on these complex-es using Ru(CO)$_2$(PH$_3$)$_3$ for the QM part. Both isomers, A and B, were optimized as local minima for each complex. The computed C-Ru-C bond angles show a neat separation between isomers A (178.2°, 174.3°) and B (136.7°, 145.6°), and are always in good agreement with available experimental data.

The computed relative energies are especially interesting. For complex Ru(CO)$_2$(PEt$_3$)$_3$, isomer B, the only one existing in the crystal, is computed to be more stable than isomer A by 3.0 kcal·mol^{-1}. The relationship between the two isomers is reversed for complex Ru(CO)$_2$(P(i-Pr)$_2$Me)$_3$. In this latter complex, A is 2.8 kcal·mol^{-1} more stable than B. Electronic effects, represented by the QM energy, always favor isomeric form B, with the two carbonyl ligands in equato-rial positions, by values of 3.1 and 1.7 kcal·mol^{-1}, respectively. Steric effects, rep-resented by the MM part, mark the difference between both species. Steric ef-fects always favor the isomeric form A, with the two carbonyl ligands in axial po-

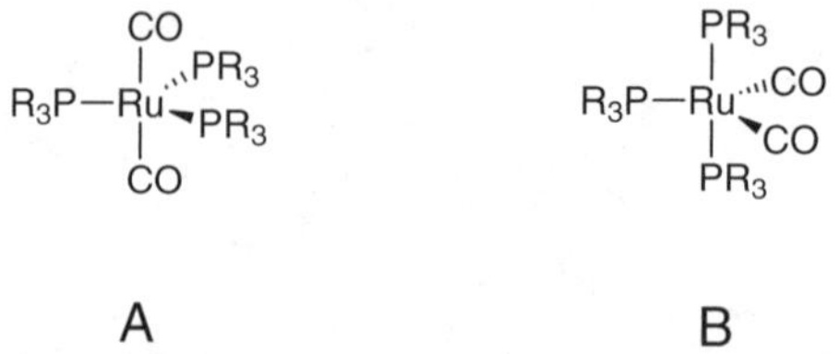

Scheme 2. The two observed isomers in Ru(CO)$_2$(PR$_3$)$_3$ complexes [52]

sitions, but they do it by quite different magnitudes. The preference is quantified as 0.1 kcal·mol^{-1} for complex Ru(CO)$_2$(PR$_3$)$_3$, and 4.5 kcal·mol^{-1} for complex Ru(CO)$_2$(P(*i*-Pr)$_2$Me)$_3$. The resulting picture is quite clear. There is an electronic preference for the placement of the π-acidic carbonyl ligands in equatorial positions (isomer B) though, quantitatively, this preference is always smaller than 5 kcal·mol^{-1} and definitely smaller than was a priori expected [54]. There is a steric preference toward the placement of the bulkier phosphine ligands in the equatorial positions (isomer A), the weight of this preference depending on the nature of the phosphine ligand. In small phosphines, like PEt$_3$, this effect is negligible, and isomer B is the more stable form. In larger phosphines, like P(*i*-Pr)$_2$Me, steric effects are strong enough to invert the small electronic preference for isomer B and isomer A becomes the more stable form.

The three structural studies that have been discussed in this subsection constitute a good example of how hybrid QM/MM methods can be applied to the study of structural problems. In the first place, they allow the calculation of geometrical structures that cannot be computed otherwise. In the second place, they allow a separate identification of electronic and steric effects. And in the third place, hybrid QM/MM methods allow a quantitative measurement of these effects in energy terms.

3.2
Olefin Polymerization Via Homogeneous Catalysis

The catalytic olefin polymerization is a reaction of much industrial interest that had been carried out traditionally with heterogeneous catalysts, namely transition metal halides with cocatalysts such as alkylaluminum compounds, following the work by Ziegler, Natta and their coworkers [55, 56]. Later on, the work by Kaminsky and coworkers [57] pointed out the large potential of early transition metal metallocenes for homogeneous catalysis [58]. More recently, Brookhart and coworkers have also shown the large efficiency of late transition metal diimine based catalysts for this reaction [59]. Scheme 3 shows the generally accepted Cossee mechanism for a typical early transition metal catalyst ZrCp$_2$Me$^+$. The key step of this mechanism is the insertion of the olefin into the M-alkyl bond [60]. Scheme 4 shows the different possible products in the polymerization of propylene.

The reason why this topic has produced a good deal of pure MM and QM/MM applications is that the nature of the products obtained is substantially affected by the presence of substituents not directly attached to the reactive center, and that play therefore an essentially steric role. In particular, these substituents affect two essential characteristics of the polymers, namely its tacticity and its weight. Propylene polymerization, for instance, can produce isotactic or syndiotactic products (Scheme 4), depending on the arrangement of methyl substituents in the carbon-carbon bond formation step. The molecular mass of the polymer, on the other hand, depends on the number of chain propagation reactions that are carried out before the chain is terminated.

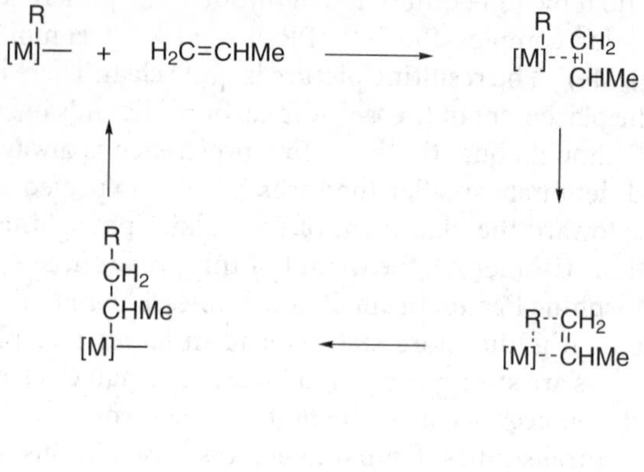

Scheme 3. The Cossee mechanism for homogeneous propylene polymerization [60]

Scheme 4. The two possible arrangements of polypropylene

The mechanism by which the tacticity of the products is controlled in early transition metal metallocene complexes has been explained mostly by theoretical calculations. In this concern, the work, mostly with pure MM methods, by the group of Corradini and Guerra [61–63] plays a central role. Their pure MM calculations share some characteristics with one-step hybrid QM/MM methods, and therefore merit some discussion here. The characteristic feature of these calculations, which have used different force fields throughout the years, is the freezing of an active core, which is assumed to be unchanged by the nature of the substituents, and the geometry optimization of the other parameters. Through these calculations they have observed that the growing alkyl chain occupies preferably the most open sector in the ligand framework, and that olefin enters the reaction complex with its substituent trans to the chain segment. For C_2-symmetric complexes identical enantiofacial olefin orientation at both coordination sites results in isotactic polymer formation; for C_s-symmetric complexes the enantiofacial orientation alternates between coordination sites and leads to syndiotactic polymers.

Similar conclusions have been obtained by Rappé and coworkers with a slightly different set of pure MM calculations [64, 65], and by Morokuma and coworkers by using one-step hybrid QM/MM methods [33, 37]. In a typical case, the QM calculations are carried out at the RHF level on the $H_2SiCp_2ZrMe^+ + H_2C=CH_2$ system, and the H substituents of the Cp rings and ethylene are replaced by the real (much bulkier) substituents in MM calculations at the MM2 level. In this way, they obtain the same control mechanism indicated above for the tacticity of the product, and they are able to compare the efficiency of catalysts containing Ti, Zr or Hf as central atoms. The fact that for this particular system the one-step hybrid QM/MM calculations essentially reproduce the results of the much cheaper pure MM calculations deserves some comment. In fact, the success of the pure MM calculations relies on the chemical skill in the choice of the guessed frozen geometries of the core on which the pure MM calculations have been carried out. Results are good because these geometries are indeed not very different from those obtained in the QM calculations. But one must keep in mind that the choice of the frozen geometries in the pure MM calculations involves a certain amount of arbitrariness. In that sense, the hybrid QM/MM calculations are more reliable.

Another closely related subject that has been recently studied theoretically with hybrid QM/MM methods is the steric control of the chain length of the polymers when Brookhart systems are applied [59]. These systems are usually Ni(II) and Pd(II) diimine based catalysts of the type (ArN=C(R)-C(R)=NAr)M-CH_3^+ type (Scheme 5). Traditionally, such late transition metal catalysts had been found to produce dimers or oligomers, the novelty of the systems recently proposed [59] being that they allow the production of polymers of high molecular weight. This process is controlled by the presence of bulky substituents on the aryl rings, and this dependence has been explained through independent QM/MM studies by the groups of Ziegler [66,67] and Morokuma [68]. In particular, they have carried out full QM/MM geometry optimizations on the (ArN=C(R)-C(R)=NAr)Ni^{II} based catalyst, with R=methyl and Ar=2,6-$C_6H_3(i$-Pr$)_2$. In Ziegler's work, the R and Ar groups were treated at the MM level with an expanded AMBER95 force field and the rest at the QM level with a non-local density functional, using a program of their own following the IMOMM prescription. Morokuma's work is done with the IMOMM program, using the MM3 force field in the MM part and the Becke3LYP density functional in the QM part. Despite these differences in the methods, the results are very similar.

One of the most significant features of these works is the comparison between the results of the QM/MM calculation with those of pure QM calculations on the

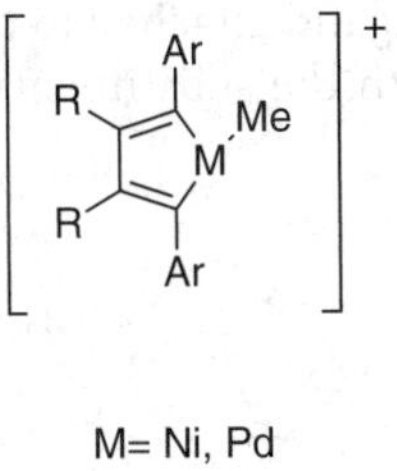

M= Ni, Pd

Scheme 5. Typical Brookhart catalyst for olefin polymerization [59]

model system [69–71]. In particular, they compare the energy barriers for three different reactions that define the nature of the resulting polymer: insertion, branching and termination. Experimental results [59] indicate that the barrier for insertion is the lowest with those for branching and termination trailing by 1.3 kcal·mol^{-1} and 5.6 kcal·mol^{-1}, respectively. The result is completely opposite in the pure QM calculation on the model system [69–71], with the barriers for branching and termination being lower than that for insertion. In contrast, the QM/MM calculation reproduces closely the experimental result, with insertion having the lowest barrier. The striking agreement between the QM/MM calculation and experiment is in itself a remarkable success of hybrid QM/MM methodology, but also has fundamental implications on the origin of the behavior of Brookhart catalysts. If termination had a lower barrier, the reaction would give rise only to oligomers of low molecular weight, which are actually the traditional products with late transition metal catalysts. This is what happens also in the QM calculation on the model system. It is therefore the steric effect of the bulky groups in the catalyst, and their subsequent destabilization of the transition state for the termination step, that is at the origin of the efficiency of Brookhart's catalyst.

The application of hybrid QM/MM methods to this particular system has been taken a step further by Ziegler's group through the introduction of ab initio molecular dynamics simulations [67]. In this way, they can shift from the enthalpy values obtained in the QM calculations to free energy values, finding that the free energy barrier for the termination reaction is 14.8 kcal·mol^{-1}, to be compared with the 18.6 kcal·mol^{-1} enthalpy obtained in the calculations mentioned above.

3.3
Asymmetric Dihydroxylation of Olefins

The osmium-catalyzed dihydroxylation of olefins is a very efficient method for the introduction of chiral centers in prochiral substrates [72, 73]. The reaction follows the scheme depicted in Scheme 6. There is an intermediate containing a 5-membered ring, and the final product is a chiral diol. The stereoselectivity is decided in the initial interaction between catalyst and substrate, leading to the intermediate containing a 5-membered ring. Because of this, all experimental and theoretical studies on the stereoselectivity have concentrated on this first part of the reaction. The NR$_3$ ligand attached to osmium is a bulky alkaloid derivative, and its interaction with the substituents of the olefin depends on the

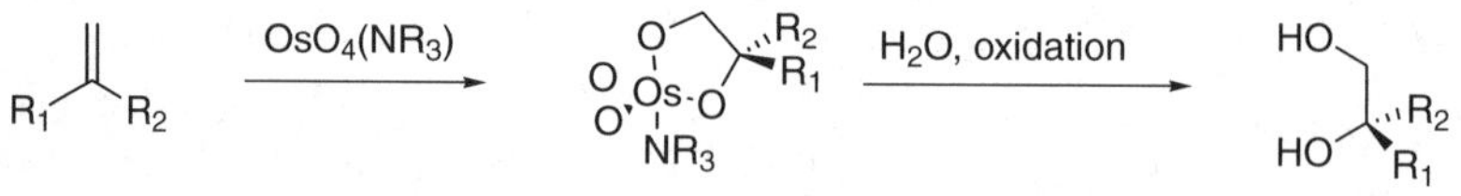

Scheme 6. Reaction scheme for asymmetric dihydroxylation of olefins with osmium catalysts

orientation of the olefin, allowing one therefore to distinguish between them. This reaction has been thoroughly studied experimentally by the groups of Sharpless [72, 74, 75] and Corey [76–78]. There has been a certain controversy on whether the mechanism takes place through a 4-membered metal-laoxoethane derivative ([2+2] mechanism, Scheme 7), but its possible existence has been finally discarded with the help of non-local DFT calculations on the $OsO_4(NH_3)+C_2H_4$ model system [79–82], which have proved that the reaction goes through a [3+2] mechanism (Scheme 7).

These calculations on the model system could not, however, address the topic of enantioselectivity, simply because in this model there is only one possible product. Several groups have analyzed the origin of chirality with pure MM and-hybrid QM/MM methods. Pure MM calculations with modified versions of the MM2 force field were carried out by the groups of Houk [83] and Sharpless [84] with the structure of the reaction center frozen. Both studies compute the correct selectivity, but their predictive power is arguable because of the arbitrariness in the choice of the frozen geometry for the reaction center. Houk and cowarkers focus on the behavior of 6-coordinate catalysts where the alkaloid is coordinated though two different nitrogen donor centers, on the assumption that this is the main reason for the increased selectivity of "dimeric" catalysts [72]. Sharpless and coworkers assume that the mechanism goes through a [2+2] mechanism involving an osmaoxoethane intermediate. In fact, although both assumptions were sound at the time the papers were published, they have since been proven wrong. This is actually a good example of the risks inherent in the

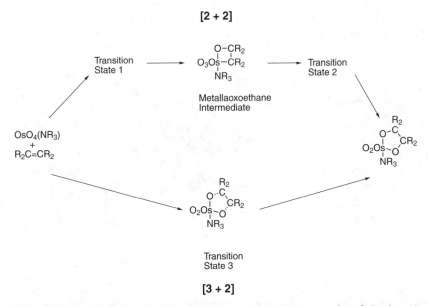

Scheme 7. The two possible mechanisms for the formation of the 5-membered ring intermediate in asymmetric dihydroxylation

application of pure MM methods to the ligands of transition metal complexes. This problem is solved by the application of QM/MM methods.

Ujaque et al. have analyzed these systems with the hybrid QM/MM IMOMM method, using the $OsO_4(NH_3)+C_2H_4$ model for the QM part. In particular, these authors have published two significant works on related complexes. The first work [85] contains a study with the IMOMM(RHF:MM3), IMOMM(MP2:MM3) and IMOMM(Becke3LYP:MM3) methods on the [OsO_4(quinuclidine)] and [OsO_4{(dimethylcarbamoyl)dihydroquinidine}] systems (Fig. 3). The paper analyzes the ability of the computational method to reproduce the structure of these two complexes, the only two catalyst analogs where X-ray diffraction data are available [86, 87]. The geometrical parameter more affected by the nature of the NR_3 group is the Os-N distance, and the discussion is therefore concentrated on its value. The experimental values are 2.37 Å and 2.49 Å for each of the complexes. The corresponding computed values are 2.429, 2.478 Å at the IMOMM(RHF:MM3) level; 2.531, 2.591 Å at the IMOMM(MP2:MM3) level; and 2.546, 2.619 Å at the IMOMM(Becke3LYP:MM3) level. These results are quite illustrative of the performance of the method. On one hand, the experimental observation of a shorter bond distance in [OsO_4(quinuclidine)] than in [OsO_4{(dimethylcarbamoyl)dihydroquinidine}] is well predicted by the calculation. On the other hand, the computed values are quite far from the experimental data, with discrepancies up to 0.18 Å. This difference is probably related to the lack of electronic effects in the IMOMM method. So, the conclusion of this study is that while IMOMM performs poorly in predicting accurately the Os-N distance, it rates highly in reproducing the steric interactions of the N substituents. Since steric interactions ought to play the main role in the stereoselectivity of the reaction, IMOMM will be a good method for the study of the asymmetric dihydroxylation process.

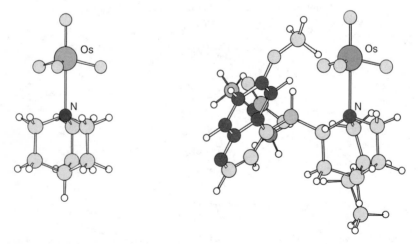

Fig. 3. IMOMM(Becke3LYP:MM3) optimized structure [85] of complexes [OsO_4(quinuclidine)] (*left*) and [OsO_4{(dimethylcarbamoyl)dihydroquinidine}] (*right*)

Quinoline A

Quinoline A

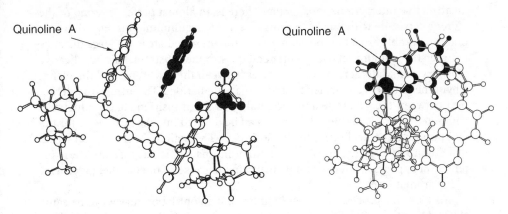

Fig. 4. Two views of the IMOMM(Becke3LYP:MM3) optimized structure [88] of the intermediate in the reaction $(DHQD)_2PYDZ \cdot OsO_4 + CH_2 = CHPh$. The styrene substrate and the OsO_4 subunit are *highlighted in black*

The second work by the same group [88] illustrates precisely this point. In this case the study is focused specifically on the $(DHQD)_2PYDZ \cdot OsO_4$ [$(DHQD)_2$ PYDZ=bis-(dihydroxyquinidine)-3,6-pyridazine]+CH_2=CHPh system. This system has been thoroughly studied from an experimental point of view by Corey and coworkers [78], having been shown to produce with high stereoselectivity the R isomer. The IMOMM(Becke3LYP:MM3) study [88] of the most likely path leading to the R isomer leads to the theoretical characterization of the reaction profile. The reaction goes through an intermediate, which is connected to the products through a transition state that is also computed. The relative energies with respect to the reactants of the intermediate, the transition state and the product are –9.7, –3.3 and –34.3 kcal·mol^{-1}, respectively. The geometries of both the transition state and the product are very similar to the results obtained in the pure MO calculation of the model system $OsO_4(NH_3)+C_2H_4$ [79], proving the validity of the studies on the model system for the late stages of the reaction. The IMOMM study nevertheless allows the location of the intermediate. Its structure is shown in Fig. 4. The presence of an intermediate had been required previously from the experimental observations of an inversion point in the Eyring plot of the reaction [89], as well of a Michaelis-Menten kinetics [78]. Very recent studies have pointed out other possible explanations for the observed kinetics [90], but the existence of the intermediate for this particular combination of catalyst and substrate seems hardly disputable.

The geometry of the intermediate (Fig. 4) has interesting implications on the ongoing discussion on the reaction mechanism. This geometry favors the U-shape proposal [78] for the conformation of the cinchona ligand instead of the L-shape arrangement [84]. The styrene substrate is sandwiched by two methoxyquinoline walls, with the parallel arrangement of the aromatic rings producing an attractive interaction. In particular, there is an almost perfect overlap of the styrene substrate with one of the rings (labeled as quinoline A in Fig. 4). A second remarkable aspect of the geometry of the intermediate is the large distance between the osmium catalytic center and the styrene substrate. This is indicative of a very weak interaction in this species between the OsO_4 unit and the olefin, a hypothesis that is confirmed by additional calculations. The attractive interaction between the substrate and the catalyst in the transition state geometry is computed to be 12.4 kcal·mol^{-1}, and it can be divided into 0.8 kcal·mol^{-1} for the QM part and 11.6 kcal·mol^{-1} for the MM part. The main part of the interaction is therefore of a steric nature. The fact that in this particular case the steric interaction appears attractive is not surprising because of the parallel placement of the aromatic rings.

Work on this process is currently under way in our laboratory with the same IMOMM method [91]. In particular, we have computed the structures of all the transition states leading to the 12 possible resulting diastereomeric products. The path discussed in the previous paragraphs, leading to the R diol product, is shown to have the lowest energy, with an energy difference of 2.65 kcal·mol^{-1} with respect to the most efficient path leading to the S diol product. Further applications will go in the direction of investigating the particular effect of different regions of the catalyst on both the rate and the selectivity of the reaction.

3.4
Agostic Complexes

An agostic complex is a molecule containing an intramolecular interaction between the metal center and a C-H bond of one of the ligands [92–94]. This type of interaction has been known for a number of years, and plays an important role as stabilizing factor of unsaturated complexes. It is also an important example of interaction between σ bonds and metal centers, with the corresponding implications in the σ bond activation process. In contrast with the topics described in the previous subsections, the relationship of agostic complexes to steric effects, and, by extension, to the application of hybrid QM/MM methods, is not straightforward. In any case, such a relationship exists, and it has been highlighted in recent publications.

The experimental observation [95] through NMR and X-ray data of two agostic interactions in $Ir(H)_2(P(t\text{-}Bu)_2Ph)_2^+$ (Fig. 5) was followed by pure Becke3LYP calculations on the $Ir(H)_2(P(Et)H_2)_2^+$ model system and hybrid QM/MM IMOMM(Becke3LYP:MM3) calculations with the same QM part on the real system [96]. The pure QM calculations indicate no agostic interactions at all (Ir-P-C angle of 118.4°), while the IMOMM calculations identify properly the agostic interactions (Ir-P-C angle of 101.7°), in agreement with experimental data (Ir-P-C, 97.0°). This result bears two important consequences. The first is that the failure of the pure QM calculations on the model system means that agostic interaction is intimately associated with the presence of the real ligands. The second consequence is that the success of the corresponding IMOMM calcula-

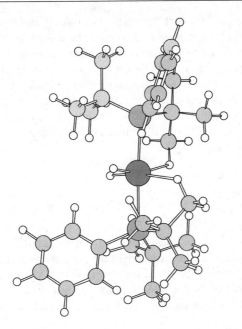

Fig. 5. IMOMM(Becke3LYP:MM3) optimized structure [96] of complex $Ir(H)_2(P(t\text{-}Bu)_2Ph)_2^+$

tions proves that the role of the ancillary substituents is of a steric nature, and that the electronic contribution is minor at best.

In a very simplistic view, it could be said that the C-H group interacts with the metal because it has nowhere else to go. However, it must be remarked that there is a real agostic interaction between the C-H bond and the iridium atom. This is shown by the elongation of the two C-H bonds closer to the metal (1.111 Å) with respect to the other ones (average of 1.094 Å).

The unexpected role played by the steric effects on agostic interactions can be viewed in a larger context. Thorpe and Ingold have identified an influence of increasing steric bulk in R of a CR_2 group on ring formation involving this CR_2: bulky R, by increasing the angle R-C-R, favors the kinetics and thermodynamics of ring formation. Shaw has extended the same idea to PR_2R' ligands [97], showing that bulky substituents R encourage ring closure, bridging and orthometallation [98]. Agostic interaction can be viewed as nothing else than an incipient form of orthometallation. Another closely related subject is the behavior of $R_2PCH_2PR_2$ ligands. When R=Ph, the ligand is most often found bridging two metals [99], but when R=t-Bu the bent monomeric two-coordinate complexes $Pt(\eta^2 (t\text{-}Bu)_2PCH_2P(t\text{-}Bu)_2)$ complexes are formed as transient intermediates, highly reactive towards oxidative addition of ordinarily unreactive bonds [100, 101]. The interpretation for this behavior is that the phenyl groups lack the bulk to enforce formation of the four-membered ring $Pt(\eta^2 Ph_2PCH_2PPh_2)$.

Therefore, it seems clear that steric effects can play an important role in the formation of agostic interactions, and, in general, in the properties of cyclomet-allated compounds. Thus, this seems to be another major topic of application of hybrid QM/MM methods to transition metal complexes for the near future. In fact, a work [102] with the same IMOMM method shows that steric effects also play a major role in the presence of agostic effects in the $Tp^*Nb(Cl)(CHMe_2)$ $(PhC\equiv CMe)$ complex, which is very different from those discussed above. At any rate, it has to be remarked that agostic interactions can also exist in the absence of steric effects [92–94].

3.5
Bioinorganic Complexes

As already mentioned in the "Introduction," biochemistry has been since the very beginning one of the main fields of application of hybrid QM/MM methods. This is obviously related to the large size of the systems involved even in the sim-plest biochemical processes. Most of the hybrid QM/MM applications have nev-ertheless been related to the introduction of solvation effects in organic environ-ments [19, 103–108], without a metal center, and therefore fall outside the scope of this review.

The work involving transition metal complexes is in a much less advanced state. In this regard, one can mention the work by Ryde, Pearlstein and coworkers, and Maseras. Ryde is studying the problem of the coordination number of zinc in the alcohol dehydrogenase enzyme. Experimental data are unclear on how many water molecules are attached to the metal center in the biological environment. The system is analyzed with a self-made QM/MM algorithm. The QM region is composed of Zn and the ligands directly attached to it, and the MM region is composed of a certain volume of the protein environment. The QM calculations are carried at the Hartree-Fock level, and the MM force field is a modification of the Merck Molecular Force Field. These calculations yield a clear preference for a 4-coordinate environment for the metal center, with the 5-coordinate isomer 95 kJ/mol above. The water molecule that could occupy the fifth coordination site in the metal center is bound instead to an amino acid of the protein.

Two QM/MM studies by different authors have been published on similar iron picket-fence porphyrin complexes [109, 110]. These species (Fig. 6) contain an iron porphyrin heme group. Apart from the 4-coordinate species where iron is coordinated only by the porphyrin ligand, the 5-coordinate species (with an ad-ditional imidazole ligand) and the 6-coordinate species (with the sixth ligand being a bent O_2 unit) are also known. The name picket-fence comes from the ge-ometry of the additional substituents attached to the porphyrin ring, which are pivalamide groups. Interest in this type of system [111] is justified because they constitute the closest synthetic X-ray model of the active sites of hemoglobin and myoglobin [112].

The study by Bersuker and coworkers [109] computes the 5-coordinate $Fe(T_{piv}PP)(imidazole)$ ($T_{piv}PP$=picket-fence porphyrin) system. It uses a partition

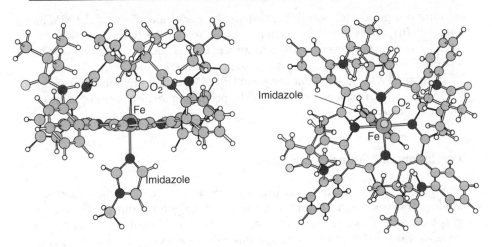

Fig. 6. IMOMM(Becke3LYP:MM3) optimized structure [110] of complex Fe(T$_{piv}$PP)(imidazole)(O$_2$)

with FeP (P=porphyrin) in the QM region, and the imidazole and the picket-fence substituents in the MM region. The particular hybrid QM/MM method is proposed by the same authors, and introduces electronic corrections from the MM atoms. The QM calculations are carried out with the semiempirical ZINDO method, and the MM calculations with the SIBYL force field. The geometrical results are in good agreement with experimental data [111]. In particular, the optimized Fe-N(porphyrin) distances are 2.10 Å, to be compared with experimental values of 2.07–2.08 Å.

The study of Maseras [110] focuses on the 6-coordinate Fe(T$_{piv}$PP)(imidazole)(O$_2$) complex, but also contains results on FeP(imidazole)(O$_2$), Fe(T$_{piv}$PP)(imidazole) and FeP(imidazole). It uses a partition with Fe(NH(CH)$_3$NH)$_2$(NHCH$_2$)(O$_2$) in the QM part. The hybrid method is IMOMM, the QM level is Becke3LYP and the force field is MM3. Geometrical parameters are also in over-all agreement with experimental data, including those involved in the coordination of dioxygen, which seems to be especially complicated to reproduce [113]. This study was also able to identify a hydrogen bond between one of the amide hydrogens of the picket-fence substituents and the oxygen which is further from the metal in the dioxygen unit.

The comparison between the approaches of Bersuker and Maseras to a similar problem is quite representative of the high diversity of QM/MM methods. The paper by Bersuker and coworkers [109] uses a large QM region and introduces electronic effects of the MM atoms in the QM region. On the other hand, it uses a relatively simple semiempirical ZINDO method for the QM calculation. The paper by Maseras [110] uses a smaller QM region, and neglects the electronic effects of the MM atoms in the QM energy. But this latter paper applies a significantly higher level for the QM calculation in the form of a non-local DFT method. Therefore, to evaluate the quality of the calculations one must take into account two points: (a) how important are the electronic effects of the MM atoms, and (b) how valid is the semiempirical method for this transition metal system. To complicate matters further, the answer to these questions can depend on the problem under study, even with the same complexes.

In summary, the application of hybrid QM/MM to bioinorganic complexes is still in a very early stage, but is one of the promising areas of development for the next future.

4
Conclusions and Future Perspectives

Hybrid QM/MM methods constitute a relatively new tool for the theoretical study of transition metal complexes. The fact that a good number of the bibliographical references cited in this chapter have been published in the last 3 years gives a measure of the fast progress in this field, progress which is likely to continue in the near future, both from the methodological point of view and from the point of view of applications.

From a methodological point of view, the large variety of different algorithms being currently used by different authors is proof of both the complexity of the issue and the vitality of the field. The IMOMM scheme [25] is appealing because of its simple robustness, and it can be the starting point for future improvements. Likely progress will be made with the introduction of electronic effects. There are already some algorithms that point in this direction [31, 114]. Specific changes in the force field for particular problems are also a promising venue [115]. Other future developments will probably come in the form of analytical computation of frequencies. In any case, the next major step towards the widespread use of these methods will probably be their introduction in standard commercial packages through user-friendly presentations.

As far as applications are concerned, the list of successes is already quite impressive. Problems where steric effects have an obvious bearing are being studied with satisfactory results, as is the case for structural problems, olefin polymerization and osmium-catalyzed dihydroxylation. New problems of this type are likely to be studied in the near future. The use of hybrid QM/MM methods has also allowed the discovery of an unexpected role played by steric effects on agostic interactions. Applications to bioinorganic systems seem also to be making a breakthrough. The near future will surely bring an expanding range of experimental problems where hybrid QM/MM methods are successfully applied.

In summary, hybrid QM/MM methods have already proved to be a powerful tool for the computational study of transition metal complexes. The next few years will probably see their consolidation as a complementary alternative to other computational tools.

Acknowledgement: Most of the writing of this chapter was carried out during the last months of my stay at the Université de Montpellier 2, France, in the position of associate researcher sponsored by the CNRS, who are duly thanked. Thanks are also due to Prof. Morokuma, Emory, for introducing me to this fascinating subject, and to Prof. Lledós, Barcelona, and Prof. Eisenstein, Montpellier, for their support during my stays at their respective departments.

References

1. Hoffmann R (1963) J Chem Phys 39: 1397
2. Albright TA, Burdett JK, Whangbo MH (1985) Orbital interactions in chemistry. Wiley, New York
3. Szabo A, Ostlund NS (1989) Modern quantum chemistry, 1st edn revised. McGraw-Hill, New York
4. Parr RG, Yang W (1989) Density functional theory of atoms and molecules. Oxford University Press, New York
5. Musaev DG, Morokuma K (1996) Adv Chem Phys 95:61
6. Koga N, Morokuma K (1991) Chem Rev 91:823
7. Børve KJ, Jensen VR, Karlsen T, Støvneng JA, Swang O (1997) J Mol Mod 3:193
8. Comba P, Hambley TW (1995) Molecular modeling of inorganic compounds. VCH, Weinheim
9. Zimmer M (1995) Chem Rev 95:2629
10. Rappé AK, Casewit CJ (1997) Molecular mechanics across chemistry. University Science Books, Sausalito, California
11. Allinger NL, Yuh YH, Lii J-H (1989) J Am Chem Soc 111:8551
12. Weiner SJ, Kollman PA, Nguyen DT, Case DA (1986) J Comput Chem 7:230
13. Brooks BR, Bruccoleri RE, Olafson BD, States DJ, Swaminathan J, Karplus M (1983) J Comput Chem 4:187
14. Cundari TR, Sisterhen LL, Stylianopoulos C (1997) Inorg Chem 36:4029
15. Cundari TR, Saunders L, Sisterhen LL (1998) J Chem Phys A 102:997
16. Warshel A, Levitt M (1976) J Mol Biol 103:227
17. Singh UC, Kollman PA (1986) J Comput Chem 7:718
18. Field MH, Bash PA, Karplus M (1990) J Comput Chem 11:700
19. Gao J (1996) Acc Chem Res 29:298
20. Gao J (1996) Rev Comput Chem 7:119
21. Bakowies D, Thiel W (1996) J Phys Chem 100:10580
22. Eichler U, Kölmel CM, Sauer J (1996) J Comput Chem 18:463
23. Eichler U, Brändle M, Sauer J (1997) J Phys Chem B 101:10035
24. Rodríguez-Santiago L, Sierka M, Branchadell V, Sodupe M, Sauer J (1998) J Am Chem Soc 120:1545
25. Maseras F, Morokuma K (1995) J Comput Chem 16:1170
26. Matsubara T, Maseras F, Koga N, Morokuma K (1996) J Phys Chem 100:2573
27. Barea G, Maseras F, Jean Y, Lledós A (1996) Inorg Chem 35:6401
28. Thery V, Rinaldi D, Rivail J-L, Maigret B, Ferenczy GG (1994) J Comput Chem 15:269
29. Tuñón I, Martins-Costa MTC, Millot C, Ruiz-López MF, Rivail J-L (1996) J Comput Chem 17:19
30. Tuñón I, Martins-Costa MTC, Millot C, Ruiz-López MF (1997) J Chem Phys 106:3633
31. Strnad M, Martins-Costa MTC, Millot C, Tuñón I, Ruiz-López MF, Rivail J-L (1997) J Chem Phys 106:3643
32. Maseras F, Koga N, Morokuma K (1994) Organometallics 13:4008
33. Yoshida T, Koga N, Morokuma K (1996) Organometallics 15:766
34. Nozaki K, Sato N, Tonomura Y, Yasutomi M, Takaya H, Hiyama T, Matsubara T, Koga N (1997) J Am Chem Soc 119:12779
35. Spellmeyer DC, Houk KN (1987) J Org Chem 52:959
36. Dorigo AE, Houk KN (1988) J Org Chem 53:1650
37. Kawamura-Kuribayashi H, Koga N, Morokuma K (1992) J Am Chem Soc 114:8687
38. Yoshida T, Koga N, Morokuma K (1995) Organometallics 14:746

39. Gusev DG, Kuhlman R, Rambo JR, Berke H, Eisenstein O, Caulton KG (1995) J Am Chem Soc 117:281
40. Maseras F, Lledós A (1995) J Chem Soc Chem Commun:443
41. Thompson MA (1995) J Chem Phys 99:4794
42. Eurenius KP, Chatfield DC, Brooks BR, Hodoscek M (1996) Int J Quantum Chem 60:1189
43. Coitiño EL, Truhlar DG, Morokuma K (1996) Chem Phys Lett 259:159
44. Noland M, Coitiño EL, Truhlar DG (1997) J Phys Chem A 101:1193
45. Coitiño EL, Truhlar DG (1997) J Phys Chem A 101:4641
46. Matsubara T, Sieber S, Morokuma K (1996) Int J Quantum Chem 60:1101
47. Froese RDJ, Morokuma K (1996) Chem Phys Letters 263:393
48. Howard JAK, Keller PA, Vogt T, Taylor AL, Dix ND, Spencer JL (1992) Acta Crystallogr B48:4338
49. Guggenberger LJ, Muetterties EL (1976) J Am Chem Soc 98:7221
50. Maseras F, Eisenstein O (1998) New J Chem 22:5
51. Aracama M, Esteruelas MA, Lahoz FJ, Lopez JA, Meyer U, Oro LA, Werner H (1991) Inorg Chem 30:288
52. Ogasawara M, Maseras F, Gallego-Planas N, Kawamura K, Ito K, Toyota K, Streib WE, Komiya S, Eisenstein O, Caulton KG (1997) Organometallics 16:1979
53. Ogasawara M, Maseras F, Gallego-Planas N, Streib WE, Eisenstein O, Caulton KG (1996) Inorg Chem 35:7468
54. Rossi AR, Hoffmann R. (1975) Inorg Chem 14:365
55. Ziegler K, Holzkamp E, Breil H, Martin H (1955) Angew Chem 67:541
56. Natta G (1956) Angew Chem 68:393
57. Anderson A, Cordes HG, Herwig J, Kaminsky W, Merk A, Mottweiler R, Sinn JH, Vollmer HJ (1976) Angew Chem Int Ed Engl 15:630
58. Brintzinger HH, Fischer D, Müllhaupt R, Rieger B, Waymouth RM (1995) Angew Chem Int Ed Engl 34:1143
59. Johnson LK, Killian CM, Brookhart M (1995) J Am Chem Soc 117:6414
60. Cossee P (1964) J Catal 3:80
61. Venditto V, Guerra G, Corradini P, Fusco R (1990) Polymer 31:530
62. Guerra G, Cavallo L, Moscardi G, Vacatello M, Corradini P (1994) J Am Chem Soc 116:2988
63. Guerra G, Longo P, Cavallo L, Corradini P, Resconi L (1997) J Am Chem Soc 119:4394
64. Rappé AK, Castonguay LA (1992) J Am Chem Soc 114:5832
65. Hart JR, Rappé AK (1993) J Am Chem Soc 115:6159
66. Deng L, Woo TK, Cavallo L, Margl PM, Ziegler T (1997) J Am Chem Soc 119:6177
67. Woo TK, Margl PM, Blöchl PE, Ziegler T (1997) J Phys Chem B 101:7878
68. Froese DJ, Musaev DG, Morokuma K (1998) J Am Chem Soc 120:1581
69. Deng L, Margl PM, Ziegler T (1997) J Am Chem Soc 119:1094
70. Musaev DG, Froese RDJ, Svensson M, Morokuma K (1997) J Am Chem Soc 119:367
71. Musaev DG, Svensson M, Morokuma K, Strömberg S, Zetterberg K, Siegbahn PEM (1997) Organometallics 16:1933
72. Kolb HC, VanNieuwenhze MS, Sharpless KB (1994) Chem Rev 94:2483
73. Lohray BB (1992) Tetrahedron Asymmetry 3:1317
74. Norrby P-O, Becker H, Sharpless KB (1996) J Am Chem Soc 118:35
75. Nelson DW, Gypser A, Ho P-T, Kolb HC, Kondo T, Kwong H-L, McGrath DV, Rubin AE, Norrby PO, Gable KP, Sharpless KB (1997) J Am Chem Soc 119:1840
76. Corey EJ, Noe MC (1993) J Am Chem Soc 115:12579
77. Corey EJ, Noe MC, Guzman-Perez AJ (1995) J Am Chem Soc 117:10817
78. Corey EJ, Noe MC (1996) J Am Chem Soc 118:319
79. Dapprich S, Ujaque G, Maseras F, Lledós A, Musaev DG, Morokuma K (1996) J Am Chem Soc 118:11660

80. Pidun U, Boehme C, Frenking G (1996) Angew Chem Int Ed Engl 35:2817
81. Torrent M, Deng L, Duran M, Sola M, Ziegler T (1997) Organometallics 16:13
82. Haller J, Strassner T, Houk KN (1997) J Am Chem Soc 119:8031
83. Wu Y-D, Wang Y, Houk KN (1992) J Am Chem Soc 57:1362
84. Norrby P-O, Kolb HC, Sharpless KB (1994) J Am Chem Soc 116:8470
85. Ujaque G, Maseras F, Lledós A (1996) Theor Chim Acta 94:67
86. Griffith WP, Skapski AC, Woode KA, Wright MJ (1978) Inorg Chim Acta 31:L413
87. Svendsen JS, Markó I, Jacobsen EN, Rao CP, Bott S, Sharpless KB (1989) J Org Chem 54:2264
88. Ujaque G, Maseras F, Lledós A (1997) J Org Chem 62:7892
89. Göbel T, Sharpless KB (1993) Angew Chem Int Ed Engl 32:1329
90. Heller D, Buschmann H, Neumann H (1997) J Mol Cat A 125:9
91. Ujaque G, Maseras F, Lledós A (1999) J Am Chem Soc 121:1317
92. Brookhart M, Green MLH (1983) J Organomet Chem 250:395
93. Brookhart M, Green MLH, Wong L (1988) Prog Inorg Chem 36:1
94. Crabtree RH, Hamilton DG (1988) Adv Organomet Chem 28:299
95. Cooper AC, Streib WE, Eisenstein O, Caulton KG (1997) J Am Chem Soc 119:9069
96. Ujaque G, Cooper AC, Maseras F, Eisenstein O, Caulton KG (1998) J Am Chem Soc 120:361
97. Shaw BL (1980) J Organomet Chem 200:307
98. Bottomley ARH, Crocker C, Shaw BL (1983) J Organomet Chem 250:617
99. Chaudret B, Delavaux B, Poilblanc R (1988) Coord Chem Rev 86:191
100. Hofmann P, Heib H, Neiteler P, Müller G, Lachmann J (1990) Angew Chem Int Ed Engl 29:880
101. Hofmann P, Unfried G (1992) Chem Ber 125:659
102. Jaffart J, Mathieu R, Etienne M, McGrady JE, Eisenstein O, Maseras F (1998) Chem Commun: 2011
103. Liu H, Müller-Plathe F, van Gunsteren WF (1996) J Mol Biol 261:454
104. Broo A, Pearl G, Zerner MC (1997) J Phys Chem A 101:2478
105. Elcock AH, Lyne PD, Mulholland AJ, Nandra A, Richards WG (1995) J Am Chem Soc 117:4706
106. Barnes JA, Williams IH (1996) J Chem Soc Chem Commun 193:
107. Ranganathan S, Gready JE (1997) J Phys Chem B 101:5614
108. Hart JC, Burton NA, Hillier IH, Harrison MJ, Jewsbury P (1997) J Chem Soc Chem Commun: 1431
109. Bersuker IB, Leong MK, Boggs JE, Pearlman RS (1997) Int J Quantum Chem 63:1051
110. Maseras F (1998) New J Chem 22:327
111. Jameson GB, Molinaro FS, Ibers JA, Collman JP, Brauman JI, Rose E, Suslick KS (1980) J Am Chem Soc 102:3224
112. Perutz MF, Fermi G, Luisi B, Shaanan B, Liddington RC (1987) Acc Chem Res 20:309
113. Loew G, Dupuis M (1996) J Am Chem Soc 118:10584
114. Humbel S, Sieber S, Morokuma K (1996) J Chem Phys 105:1959
115. Ujaque G, Maseras F, Eisenstein O (1997) Theor Chem Acc 96:146

Author Index Volumes 1–4

Printing: Mercedesdruck, Berlin
Binding: Buchbinderei Lüderitz & Bauer, Berlin